GRASSLAND SMALLHOLDING

Acknowledgements.

The authors wish to thank:

Dr. David Corke for comments about Christmas trees.
Dr. Audrey Yelland for the suggestion about blueberries.
Annabel, Doris, and Tim for comments about their experiences with livestock.
Carol for reading the manuscript.
And staff of ADAS and the Potato Marketing Board for prompt response to telephone inquiries.

GRASSLAND SMALLHOLDING

Dr. Derek Bryce and
Dr. Arabella Wagenaar

Photographs of cattle and poultry by **Harold Richards**

ISBN 0947992030

Copyright © Derek Bryce 1985. All rights reserved.

First published in Great Britain in 1985 by
Llanerch Enterprises, Llanerch, Felinfach, Lampeter, Dyfed.

ISBN 0947992030

Printed by Cambrian News, Aberystwyth.

PREFACE

We have written this little book to provide the kind of advice and information we would have wished for ourselves when each of us moved from urban professional living to farming and rural life. Much of the information and advice in this book is based on our personal experiences, and those of our friends and neighbours. We have written with United Kingdom readers in mind and have included references to sources of information about grants, premiums, subsidies, etc. but without giving specific details because the situation with regard to these things may change at any time. Smallholders must therefore check their own eligibility and their work specifications beforehand. The question of voluntary registration for VAT has been avoided; although advantageous to the full-time farmer, it may not always be so for smallholders engaged in some additional commercial activity. This is something to be considered in the light of each one's personal circumstances and proposed activities.

We have intentionally stressed economic aspects of farming, for an enthusiastic but uncommercial approach can easily result in financial loss on an unexpected scale. Some of the suggestions we offer, are, of course, only intended as food for thought, and the would-be potato grower, asparagus producer, etc. should read up the subject in greater detail, carry out market research, and seek expert advice before embarking on a major project. We have included diagrams of some of the useful things, sheds, brooders, feed boxes, etc. that we have constructed on the farm, giving enough detail to satisfy the intelligent do-it-yourself enthusiast.

We have tried throughout to be concise and informative, without wandering off the subject. We have therefore said nothing about nature conservation, although we would hope that most smallholders would wish to retain as much natural beauty as possible on their land, and perhaps keep a few odd corners undisturbed for the wildlife.

We hope this book will provide a useful guide for urban dwellers who are intending to take up farming, and those who have already done so. Good farming.

Derek Bryce
Arabella Wagenaar
Wales, March 1985

LIST OF ILLUSTRATIONS.

CONTENTS

CHAPTER 1. MAKING A START.

Back to the land.

During the past two decades or so, more and more people living in urban areas have become interested in moving back to the land. A considerable volume, of mostly well-produced literature, on the theme of self-sufficiency, has been produced to cater for this interest. Many of those who buy smallholdings are seeking not only interest and enjoyment, but also a tangible return from their farming activities - either home-produced food, cash, or both. Some, who buy holdings, seek to supplement a small pension or investment income; others, often younger people, invest all their capital in a holding with the intention of making a living from it. Many succeed; some do not. Smallholdings and small farms do not have sufficient acreage to be viable on grass production alone; some additional income is needed, either from intensive meat production, vegetable or flower growing, or from some other source. Most of those who succeed find they prefer their new way of life even when they are working harder, and for less money, than ever before. Lack of success is sometimes due to having started off with an unrealistic approach, but sometimes those who sell their holdings do so because they have realized that some other business would suit their needs better. Many of those who sell their holdings still stay in the rural environment; few return to the cities. Many of those who sell, realize more money than they started with, for land generally increases in value. A very few of those who sell their holdings, do so because they find farming not to their taste.

Do you really want a smallholding?

Compared with urban dwelling, farming can provide a highly-rewarding and satisfying way of life. Living in the rural community can also be most enjoyable. To succeed, it is necessary to start off with a realistic approach to farming, and to be sure that you are 'right' for farming (or vice-versa). This book has been written with the aim of providing an insight into what practical farming entails; to find out if farming suits you, it is a good idea to attend practical courses provided by Colleges of Agriculture, or others, and to spend some working weekends and holidays on farms, or properly-run smallholdings.

Before buying a smallholding, it is important to take a hard look at yourself, especially your motives, needs, and expectations.

You may be motivated from a long-standing interest in farming, by a desire to find a 'better' or 'more satisfying' way of life (however you define these terms), or you may be motivated from a fear of the eventual collapse of civilization obliging all to return to a subsistence economy. If you have a genuine interest in farming, all well and good; if you are looking for a better way of life, farming may be the answer, but perhaps a market garden, village shop or post office, or a country pub would suit your needs better? You should weigh up the pros and cons of each possibility. If you are motivated from a fear of what might happen in the future, this alone will not guarantee that you will be good at farming, and you should satisfy yourself first by some practical experience. If part of your motive is to get away from difficult personal relationships, remember that you will be taking yourself with you when you move, and you may create similar situations in your new environment.

You may think that your needs, psychological and financial, are few and simple; that you will adapt to isolation, financial restraint, and working seven days a week, once you have moved. You should discuss this with friends, for our friends often know aspects of ourselves that we are not fully aware of, and can sometimes tell us if we are deceiving ourselves. You should note also that, should you fall into dire straits, ownership of land may disqualify you from Social Security benefits, and you may be obliged to sell up!

Some financial considerations.

In relation to capital invested, the return for working the land is considerably less than in many other businesses and industries. Some farms, in fact, lose money, sometimes through poor management, sometimes because they give insufficient return to cover high capital investments. The financial advantage of owning farm land comes from its keeping pace with, or exceeding, inflation. Farming gives therefore a relatively small annual income, with a capital gain at the end, realizable only when the land is sold, (and subject to capital gains tax if you are giving up farming). If you have children, and your family income is small, you may be able to claim Family Income Supplement.

Real, and fantasy farming.

For farming to pay at all, the livestock must be commercial breeds kept under economic conditions; any other saleable products must be in sufficient quantities to make their marketing worth while. Uncommercial livestock often cost as much to feed as their commercial counterparts, and they provide little return at the end of the day. The transport costs of taking very small quantities of assorted surplus produce to town or market may take most of the profit away. Enthusiasm can be a smallholder's greatest asset, but over-enthusiasm, by which one becomes carried away into fantasy, can be a recipe for disaster. Those who imagine that they will be able to produce their own meat and vegetables, harvest their wheat plot, grind their own flour and bake their own bread, milk their cow and make butter, cheese and yoghurt, clip the wool off their sheep, spin, knit and weave it into their own garments, cook on home-produced methane gas, and the rest..... and live happily ever after, are probably lost in the realms of fantasy farming. They should remember the old saying: 'he who tries to do too many things, even easy things, will not succeed.' Perhaps someone, some-where, manages to do all of these things; in our experience, most smallholders find time to do only a few of the more practicable of them. In the ancient past, there may have been times when people built their own houses, and produced their own food and clothing. The people of those times had a much simpler outlook, and fewer needs, than modern man. In recent historic times, when many remote farms functioned at an almost self-sufficient subsistence level, families were much larger, providing an adequate labour force for the wide range of activities that were carried out.

If you have come this far, and still want a smallholding, read on!

Buying a holding.

The selling price of the majority of farms is governed by the acreage, the nature of the farm house having only a limited in-fluence. The price per acre of farm land depends on the state of the agricultural industry at the time, the geographical location, the quality of the land, and the influence of land investors and spec-ulators.

The price of smallholdings, on the other hand, is influenced by the pressure of demand, and by the type of house. In the London commuter belt, and some of the more beautiful and favoured parts of the country (including some of the National Parks), smallholdings are very expensive indeed. Elsewhere, often on the poorer soils and in hill country, the price of the average holding seems to be linked to the selling price of suburban city houses, for this represents the capital that most would-be smallholders can raise. There are few smallholdings on the fertile plains and valley floors; former small farms and holdings having been 'swallowed up' by larger farms in those parts.

Once you have decided to buy a holding, you will probably have decided on an area, or areas, either from personal preference, or because holdings there are likely to be within your price range. Apart from looking in periodicals such as the 'Exchange and Mart,' the first thing to do is to contact local estate agents, and also to arrange to have local papers sent to you by post. If you are looking for a cheap, run-down holding, perhaps with only a derelict cottage on it, you would do well to spend as much time as possible in the area, perhaps using a motor-caravan. Estate agent's commission on cheap properties is relatively small, so they tend to advertise them once or twice in the local paper, but seldom consider it worth their while mailing details to clients at a distance. Furthermore, cheap properties are often 'snapped up' within days, sometimes by a builder or speculator. Sometimes cheap properties can be found by asking around in village shops, pubs, etc. If you do buy a cheap derelict property, you should not only check on planning permission for the property itself, but also for a caravan that you can live in whilst the property is being restored.

When you are on your way to look at a property, remember to be polite and pleasant with everyone you meet; you never know who may become your future neighbour! When viewing, remember to pay as much attention to the land, as the house. Many urban dwellers inspect the house carefully, and only cast a cursory glance over the land, sometimes with expensive consequences later. When looking over the land, note if it is facing south, or north, check that the pasture is well-drained and in good heart, that fences, posts and gates are in good order, and that water is available in all the fields. Check also the size and condition of the outbuildings, for they are expensive to extend, or renovate. It may be perfectly in order for you to buy a cheap holding with both house and land in poor condition, provided that you have considered fully what this entails. Apart from the cost of house and land improvements, you will need capital on which to live whilst they are carried out, and the holding is 'got on its feet.' Above all, avoid becoming so carried away that you adopt the attitude of 'any old house, any bit of land, and somehow, someway, we'll make a go of it' - this can be a formula for disaster. If you are set on the idea of farm-gate selling of some of your produce, don't buy a remote holding; it must be on a reasonably busy road!

Once you are happy about a holding, you may put in an offer, assuming you are in a position to raise the finance. If the holding is being sold at auction, remember that if your bid is accepted, you have legally entered into a contract, and should complete within twenty-eight days. When bidding at auction, try to keep a poker-face, and avoid starting to bid too early. If you really want a property, wait until the other bidders start to waver - if the bids have been coming in steps of say, £1000, a bidder may ask if he may go up, say, £500 - that is the time to put in a bid of £1000 more, and this may discourage those who are wavering. For the rest, be self-controlled and don't bid more than you can afford, or more than the property is worth.

Moving in.

One of the first things to do, once you have settled in, is to check the perimeter fences or hedges, and to take whatever action may be necessary to make them stock-proof, for 'good fences make good neighbours.' It is also a good idea at this stage to go round and introduce yourself to your new neighbours - they may tell you things about your land, and its past uses, that could be useful to you when planning your farming activities. For example, they may tell you that in the past a particular field was used successfully for growing potatoes, or that a previous owner tried a particular breed of sheep, without success, and so on. When talking to neighbours, remember that country ways are different from town ways. If you listen and 'fit in,' you will find the country people helpful and kindly. Remember that they have been farming a lot longer than you, so whatever you say, don't criticize.

If you have not already done so, you should obtain the free booklet 'At the Farmer's Service.*' This tells you about grants, subsidies, and regulations. If you are intending to carry out a lot of improvements to your holding, there may be grants for such things as new fences, drains, outside stock-pens, etc. To qualify for grant-aid, you must carry out the work to the correct specification, and your holding must be capable of providing the greater part of the employment of at least one reasonably skilled person. Subsidies may also be available under the Hill Livestock Compensatory Allowances Regulations if your holding is of more than three hectares (about seven acres) and it is in a less-favoured area. Grants for the erection of outbuildings are generally not available to smallholders. If you are going to keep livestock, you should apply to the local office of the Ministry of Agriculture, Fisheries and Food for a holding number, and you should keep an Animal Movement's Record Book in which you record details of livestock purchased and sold. If you are intending to keep one or more cows, you should also apply for a herd number and have serial ear tags prepared with this number on them, for tagging new-born calves.

*In England and Wales write to: MAAF (publications), Lion House, Willowburn Estate, Alnwick, Northumberland, NE662PF. For equivalent information in Northern Ireland, write to DNAI, Dundonald House, Newtownwards Road, Belfast BT43SB. In Scotland write to: Intelligence Branch, DFAS, Chesser House, Georgie Road, Edinburgh, EH113AW.

CHAPTER 2. MANAGING THE LAND.

Most British farm land would, under natural conditions, be decid-
uous woodland. The green fields that we see everywhere today, are
prevented from reverting to scrub and woodland through grazing,
and other activities of land management. With the exception of
our uplands, and some ancient water-meadows, most of today's
grassland is not even composed of native grasses; it has been
ploughed and re-seeded with modern varieties.

The term 'grassland' is, in fact, a misnomer, since most pasture and
hay fields contain also a variety of other plants; only on silage
grounds, where a high nitrogen fertilizer regime has been used, do
we find almost pure stands of grass, often rye-grass. Other plants
which occur on pasture include, from the farmer's point of view,
beneficial species, and unwanted weeds. Pre-eminent amongst the
former are the clovers which have the ability to 'fix' atmospheric
nitrogen, providing 'free' nitrogenous fertilizer. Other common
plants include the ubiquitous daisy, plantain, and yarrow. Some of
the plants that occur amongst the grass are deep rooting, bringing
up nutrients, minerals, and trace elements from deeper down. They
can be beneficial therefore to grazing stock. Most of the seeds sold
for re-seeding pasture are, in fact, mixtures of grass and clover;
some seedsmen also sell mixtures of herb seeds, such as chicory,
sheep's parsley, ribgrass, and yarrow, for adding to the grass-
clover mixtures. Notorious amongst the unwanted species are
buttercups, thistles (especially the creeping kind), docks, ragwort,
bracken, reeds, and mosses. Clumps of nettles, generally unwanted
nowadays, used to be harvested young and cooked like spinach,
made into nettle beer, or scythed and made into a palatable hay.

If you buy grassland in poor condition, you may opt either for a
programme of grassland improvement, or for re-seeding. The
decision as to what action to take depends partly on how bad a
condition the grass is in, and on expense.

Grassland improvement.

There are many ways in which improvement may be carried out.
The decision as to what action to take depends on the nature of the
problem. We list below a variety of ways by which improvement
may be carried out.

Burning. Sometimes fields have been neglected and undergrazed.
This happens quite often if the stock were sold off quite some time
before a holding was sold. There will be clumps and mats of old
dead or dying grass which will prevent new shoots from coming

through in the spring. A burning oily rag can be dragged back and forth over the field to burn off all the clumps and tussocks. Quite often this treatment can give a field a new lease of life. When burning, always work downhill and into the wind as far as possible, for this helps to control the fire. There are regulations governing when, and under what conditions, you may carry out burning; you should contact the local Council for information.

Liming. Grass grows best in slightly acid soils, at a pH* of around 6 - 6.5. Over the years, many soils gradually become more acid, the pH decreasing. As the soil becomes more acid, clover declines, sometimes replaced by bird's foot trefoil; mosses and rushes may appear, and grass production drops. When the pH falls below 5.5, the land may be limed with a view to increasing grass production, restoring the clover, and eliminating unwanted acid-loving weeds. The degree of improvement, however, will depend on the age of the grass ley; if it has been re-seeded with a modern grass-seed mixture, it should show quite a significant improvement; if it is an old ley, it may have been invaded by native species of acid-tolerant grasses, and the improvement will not be so marked. Lime may be applied as burnt lime, or ground limestone. The former is more expensive, but quicker acting; the latter is slowly dissolved by the acids in rain water. The quantities to be applied depend on the pH, and the type of soil. Application of excess lime is not beneficial, since grass grows best in slightly acid soil; excess lime can also reduce the mobility of mineral phosphates. Lime is not a fertilizer; it simply helps to maintain soil at the optimum pH for grass growth.

Slagging. Years ago, basic slag, containing about fourteen percent phosphate, was spread at about half a ton per acre, to last for about three years. On poor quality, upland pastures this was often the only fertilizer applied, sometimes following an application of lime. Nowadays there is not so much basic slag on the market, and many farmers use mineral rock phosphate such as Gafsa, or synthetic triple superphosphate. Gafsa contains about twenty-seven to twenty-nine percent phosphate, so less of it need be spread than basic slag; it is slowly dissolved by the acids in the soil and therefore it does not work immediately after liming. Gafsa works best on slightly acid soils; it is not suitable for use on calcareous soils. If the land has been limed, Gafsa should begin to work about six months later. Triple superphosphate is usually around forty-five to forty-seven percent phosphate, so it can be applied in even smaller quantities; it is expensive and does not contain the

*PH is a measure of alkalinity and acidity; pH 7 is neutral; values above 7 are alkaline, values below 7 are acid.

trace elements that occur in natural mineral fertilizers; it has the advantage of being soluble and quick acting. Although phosphatic fertilizers are often applied following liming, in programmes of grassland improvement, application of basic slag or mineral phosphate alone will often improve a pasture.

Mowing. Some unwanted plants, such as rushes and bracken, cannot withstand repeated cutting. They may be scythed, or cut with a tractor-mounted finger-bar mower, or flail mower, set to cut several centimetres above the ground, leaving most of the grass. This is a slow process, involving several mowings a year over at least a two-year period. Mowing can be used to control thistles, preventing them from flowering, but it does not eliminate them.

Using herbicides. Herbicides may be selective, or non-selective. Some of the older selective herbicides were selective against broad-leaved plants, as compared with narrow-leaved grasses; therefore they killed useful herbs as well as weeds. Many modern selective herbicides have been developed to be more toxic to certain groups of plants, than to others; at the correct dilution they may be sprayed to kill an unwanted plant, for example bracken, without too much damage to the underlying vegetation. It is important to obtain the correct herbicides for the particular kinds of weeds you wish to kill. Non-selective herbicides can be sprayed wholesale to kill all the vegetation in a field, prior to re-seeding, without deep ploughing. These non-selective herbicides can, however, be used selectively by devising ways of targeting only the unwanted plants. One such method is called weed-wiping. A tractor-operated weed-wiper relies on the fact that many weeds grow taller than the grass. The wiper holds a horizontal rope wick, impregnated with herbicide solution, behind the tractor. The height of the wiper is adjusted so that as the tractor moves along it wipes the tops of the weeds without touching most of the grass. This method can be used for killing docks, nettles, and other tall-growing weeds. The operation may have to be repeated since small plants are not touched, and grow up later. The smallholder can use a hand operated weed-wiper. This consists of a long tube holding the herbicide solution, with a short rope wick fixed across a T-piece at the lower end. For small scale operations there are special gloves that can be impregnated, and whose touch is deadly. Plants wiped do not die instantly - it takes several days before any symptoms show. From the point of view of economy and environmental hazard, weed-wiping has a lot of advantages over wholesale spraying. Take great care when handling undiluted herbicides; make sure they are diluted correctly, and wear a breathing mask if the manufacturer recommends it.

Ditching. Wet fields can sometimes be improved just by clearing blockages in existing ditches and drains. If a field needs all the ditches digging afresh, it is a good idea to hire a mechanical digger.

Underploughing. A tractor-operated underplough cuts a horizontal slice through the soil beneath the grass roots. Underploughing effectively cuts a horizontal slice right across a field; sometimes it makes quite an improvement to the drainage. It is much cheaper than putting in a new drainage system.

Moleing. A mole-plough has a large bullet, or bomb-shaped piece of steel which is dragged underground, leaving a tunnel behind as it goes. When drains are laid in a field, a herring-bone pattern of moles is made to lead underground water to them; eventually the moles fill in, and new ones have to be made. If you can find a plan of the drains in your field, you can lay moles in the correct pattern - if not, mole down the slopes and towards the ditches at the edge of the field.

Laying drains. If the above possibilities seem inadequate, it may be necessary to lay a complete drainage system. This involves digging a series of trenches in the field and laying down ceramic, or perforated plastic drains, with an infilling of gravel above them. Earth is replaced above the gravel, any surplus being evenly distributed over the field. An appropriate pattern of mole drains has to be made to service the fixed drains, and measures have to be taken to drain the water away from the field, via ditches or wide drains, to the nearest watercourse. This is an expensive procedure. Before embarking on it, you should consider possible alternatives, such as practising strategic grazing, having your stock on the wettest fields in dry weather, and on the driest pastures in wet weather.

Re-seeding.

Re-seeding is often carried out by ploughing, and then repeatedly disc harrowing until a fine tilth has been obtained for a seed bed. Stones often need to be removed during this process. Lime, if required, and fertilizer should be spread just before the final disc harrowing. The seeds may then be drilled, scattered mechanically or by hand, and rolled in. Rolling will also push back any protruding stones. For hand sowing, there are special devices which help to give an even coverage; if you simply scatter the seeds, mix some sawdust with them so that you can see where you have sown.

Some farmers nowadays kill off the vegetation with a herbicide, and shallow rotavate the ground to provide a seed bed. Some, after using herbicide, burn off the dead vegetation and then direct-drill the seeds. Some claim that these methods are cheaper, and do less harm to the earthworm population than ploughing does. Partial re-seeding may be carried out using a special drill which lays down stripes of herbicide in which seeds are drilled at the same time. This method is not suitable for fields containing creeping varieties of grasses.

A couple of years ago one of us had cleared about an acre of scrub - mostly blackthorn and gorse. The ground quickly became over-grown by brambles, bracken, and rose-bay willow-herb. These were killed off by spraying herbicide from a knapsack spray. The dead plants were scythed and raked together into bonfire piles. In the autumn a mixture of grass and clover seeds were scattered by hand on the exposed ground. A good covering of grass was obtained by the following spring. This method of scattering seeds on bare ground would probably not be successful in spring - the birds would eat them; in autumn the birds have plenty of other food available.

Grass seedlings are susceptible to drought. Sometimes the seeds of spring-sown grass are mixed with a small amount of cereal or rape which will act as a nurse crop, helping to protect the grass seed-lings from dessication. Sometimes, when cereals or rape are sown as full crops, they are undersown with a grass-seed mixture to provide grazing after the crop has been harvested or grazed off.

Some farmers prefer to re-seed in spring, because this provides grazing later in the year. Spring-sown grass is often accompanied by weeds; the latter are not such a problem with autumn sowings. It is a question of swings and roundabouts.

Working the land.

There are a number of routine operations to be carried out every year. In early spring the grass may be chain harrowed, rolled, and fertilized. Later on there may be topping to carry out, and all the activities associated with conservation - hay, or silage making.

Chain harrowing. The chain harrow opens up any clumps, mats, or tussocks of old grass, permitting new shoots to come through. It helps to get air down to the grass roots, and it spreads out mole hills (if the latter have not already been raked by hand). Chain harrowing is not always essential - it depends a great deal on the condition of the grass.

Rolling. A heavy roller will push protruding stones back into the soil; this is important on the conservation grounds, for protruding stones can break the blades of the mowing equipment. Like chain harrowing, this operation is not always essential; it depends on the condition of the land.

Fertilizing. Your fertilizing policy will depend on whether you decide to be an organic farmer, or not.

Organic fertilizing. You can spread muck direct from your barn, or after it has been composted. If you keep poultry, your poultry manure should first be composted. People used to mix superphosphate of lime with composting poultry manure, but the latter is expensive nowadays. All manure tends to depress the pH of land (makes it more acid) when it is spread. It could be useful to mix some agricultural lime in with the composting manure, to help counteract acidity. If you make compost in the open, you will lose nitrogen and soluble organic compounds through leaching, unless you add a slurry pit to the drain from your compost heap. Slurry can be spread as an organic fertilizer, but few smallholdings have slurry pits; the latter are more a feature of large dairy farms. A roofed-over compost pit will greatly reduce losses through leaching.

Commercial fertilizers that may be considered for organic systems of farming include dried seaweed, such as calcified Cornish seaweed, a basic phosphatic fertilizer, and 'natural' mineral fertilizers such as Gafsa and other rock phosphates, and muriate of potash.

This means that you will be able to provide as much phosphate and potash as is needed, plus humus and some nitrogen from muck or compost. If you in-winter some, or all of your stock, without the benefit of a slurry pit, you will lose the nitrogen from their urine. Furthermore, each time you sell an animal, you will, in a sense, incur a nitrogen deficit. If you wish to sell your produce as 'organic' you will have to try to compensate by encouraging the growth of clover which will add 'free' nitrogen to your land, and perhaps also by growing green manure crops, and ploughing them in. This is fine if you are growing some cash crops, for they may fetch a higher price with an authentic organic label. If you are raising stock, however, and you simply have a leaning towards organic farming, rather than an immovable fixation, you might consider using a small amount of synthetic nitrogen to compensate for inevitable losses of 'natural' nitrogen. We would also point out that when an animal's 'natural' urine waters the ground, it results in urea, ammonia, and nitrates which are chemically the same as the constituents of synthetic nitrogenous fertilizers. Our reason for making this point is that few abattoirs or butchers will give a higher price, or show special interest in, 'organic' lambs or beef cattle.

Combined organic and inorganic fertilizing. Most farmers and small-holders spread whatever muck, compost, or slurry they collect, and in this sense they are part-organic. Most farmers and small-holders also spread fertilizers, including synthetic phosphatic and nitrogenous ones. If you don't object to using these products, the question then becomes one of finding the right balance, and avoiding excesses.

The following comments may help you to understand how to work out a fertilizer regime. For maximum growth, grass requires some phosphate, together with larger and almost equal amounts of potash and nitrogen. The correct proportions of phosphate, potash and nitrogen, for maximum growth, can only be used on conservation grounds. On pasture there is too great a risk of grass staggers (hypomagnesaemia) if a high potash regime is used, for the grass grows so quickly that it fails to take up sufficient magnesium to satisfy the needs of the grazing animals. Therefore a compromise has to be worked out for grazing, between the ideal of of maximum grass production, and avoidance of the risk of magnesium deficiency in grazing stock. Sheep tend to deplete a pasture of phosphate, much more so than cattle do. Our conclusions based on the fore-going are:

1. On sheep pasture to apply about twice as much phosphate as potash. For example 80 units* of phosphate and 40 units of potash may be applied in autumn and winter to last for two or three years. Small amounts of nitrogen may be applied, in the form of urea or nitrate, as required. We personally favour applying nitrogen on part of the land in early spring to provide an 'early bite,' and likewise in autumn to provide some cool-stored grass for late autumn and early winter, relying on residual nitrogen from clover for the main growing season.

2. On cattle pasture to apply equal amounts of phosphate and potash. 70 - 80 units of each may be put down in autumn or winter, to last for two years. Nitrogen to be applied as in (1) above, or in moderation, as required.

3. On conservation grounds to apply muck and slurry when the fields are shut off to the stock. For hay, to apply up to 50 units of potash and 30 of phosphate, preferably during the winter, and up to 50 units of nitrogen when it is warm enough and the grass is growing. Too much fertilizer on a hay ground can give a crop so thick that it is difficult to dry. More may be used for silage.

*Units per acre. The old fertilizer unit was strictly one hundredth of a cwt. Thus a 50 kg. bag of a 40% fertilizer spread on an acre will represent about 40 units.

Phosphate and potash are not leached from most soils by rain, but nitrogen is. That is why the former may be applied during the autumn or winter. The idea of putting all of the muck on the conservation ground is that it avoids contamination of the grazing pasture. An exception should be made if any of the grazing land is deficient in humus. The above comments are offered as a guide only. You may, of course, prefer to use compound fertilizers, instead of the 'straights' recommended above. Generally speaking, compounds work out a little more expensive, unit for unit. If you opt for compounds, be sure to select those which fulfil your needs. There are compounds on the market specially formulated for conservation. Popular compounds, such as 20.10.10 and 29.5.5 have much more nitrogen in them than potash; using them on conservation grounds could therefore be wasteful, unless you have spread enough muck or slurry to provide additional potash. These popular compounds are, however, passable on general grazing land. If you are uncertain which fertilizer to use, you may first have your soil analysed - some fertilizer firms will do this free of charge. Some farmers during the last few decades have opted for high nitrogen regimes, thereby keeping their land heavily stocked. The smallholder may be tempted to do likewise, in order to keep more animals on a small acreage. We would suggest that a good alternative would be to keep fewer animals of better quality, rather than risk the health hazards sometimes associated with high stocking densities, and possible long term deterioration of the soil.

Spreading. A tractor-driven muck spreader will distribute your muck for you; otherwise you have to fork or shovel your muck or compost in dollops from a trailer, and then rake it out by hand. On a small acreage, fertilizer may be spread by hand; powdered fertilizers are unpleasant to spread this way and you should wear a breathing mask; granular fertilizers are easier to spread - carry the fertilizer in a bucket and flick it in small amounts using a shallow tin dish. A tractor-driven fertilizer spreader is an asset if you have to spread more than a few acres. It will have an adjusting device so that you can control the amount spread per acre. Tractor-mounted spreaders operate from the tractor's power take off. There are also spreaders which can be towed behind a vehicle, obtaining their motive spreading force from the revolutions of their own wheels.

Topping. Later in the growing season, some of the grass on the grazing land may flower and go to seed. This may be cut using a finger-bar, or flail mower, set to 'top' the flower and seed heads, without cutting the rest of the grass. This operation is not essential but it does help to keep the grass growing. Topping may be used also to prevent thistles from flowering.

Hay-making. The best time to cut grass for hay is when it is flowering, before it has gone to seed. The British climate does not always permit cutting at this time. The quality of hay depends not only on when it was cut, but on the conditions under which it was made, and stored. Hay which has been made with a lot of bashing about under a hot sun will have suffered leaf shatter - the blade part of the leaf will have disintegrated into powder, leaving the midrib. Likewise hay made from grass cut when it has gone to seed has a high proportion of stem to leaf. In either case the hay is high in fibre, but relatively deficient in protein. Hay which is stored at too high a moisture content may go mouldy. One of the common moulds, Aspergillus, can cause farmer's lung.

In the old days when hay was made largely by hand, large families lived on farms and there were plenty of village children to help. Your smallholding will probably be a long way from your family and friends. If you insist on making hay the old way, you may find your neighbours unwilling to help; they will almost certainly do theirs by machine and think it crazy to go back to the old ways. You may find also that the village children are too distracted by T.V., computer games, cycles and the like, to want to help you. One of us made two acres of hay the old way, almost single handed - he decided that once was enough, and now uses a local contractor to do the job by machine. If, despite everything, you decide to do it by hand, remember that the best quality hay is wind dried, rather than sun dried, for the former avoids leaf shatter. Instead of the old way of rowing up the hay, turning and tedding it, why not try using tripods? First the hay must be mowed and allowed to wilt overnight on the ground; then it should be turned over using hay rakes to allow the part that was underneath to wilt. Whilst this is going on, numbers of tripods should be erected, made from lengths of hazel or ash coppice, or similar, wired, or lashed together with twine (fig. 2-1). Your field will soon look like a half-made Red Indian village. The hay should be draped over the horizontal bars of the tripods, taking care to leave holes for the wind to get in - a bent piece of corrugated sheeting can be used temporarily, and drawn out when the tripod is finished. Continue building up the tripod until it looks something like a busby hat. Proper ventilation is essential, as is wilted grass - if you put too much wet, green grass on a tripod, it may go mouldy. Once finished the tripod will turn some rain. Your hay will have to be stored inside, loose, or made into a traditional haystack. Other methods of wind drying include draping the grass over fences or temporary 'clothes lines', or sticking it high up on poles.

There is a lot to be said for making hay with modern machinery. The British climate seldom gives long periods without rain, and mechanical hay-making can be much quicker. Furthermore, baled hay is a saleable, and often profitable commodity. Loose hay is more difficult to sell and transport. There are two approaches to modern hay-making. Some people, after mowing, allow the grass to wilt in rows on the ground, turning it over the next day using an 'acrobat' of spiked wheels set at right angles to the rear of the tractor. They let it wilt again a few hours, and then use a tedder which whizzes the grass about and fluffs it up so that air and wind get into it. This process is repeated until the hay is ready for baling. It is ready when a bunch of hay strongly twisted between the hands no longer exudes moisture. The time depends on the weather, day length, and the density of the crop. When it is ready for baling it is rowed up, using an 'acrobat' or equivalent, and then baled. Others stir and ted the grass almost immediately after mowing, repeating the operation daily until it is ready for baling. Once hay has been baled it can be loaded on to trailers and quickly stored under cover. Bales stood on their ends can turn a little rain, but not much. Some farmers bale their hay a little early and complete the drying process under cover. The simplest type of barn bale-dryer uses tunnels made of wooden frames which run under the stacks of baled hay. Fans are set up to blow air into the tunnels.

Silage-making. Few smallholders go to the trouble and expense of constructing a silage pit. If you decide to make silage, the grass should be cut earlier than for hay, at ear emergence. The best quality silage is made by machines which chop the grass into short lengths. These multi-chop forage harvesters are expensive; most smallholders making silage this way would have to think of paying a contractor. The grass is best wilted on the ground for about twenty-four hours. This reduces the moisture content and helps to improve the quality of the silage. For successful silage making, air must be excluded as soon as possible, by sealing the plastic lining of the pit all round, and at the top - this permits rapid build up of carbon dioxide and the onset of acid conditions, for silage is effectively pickled grass. An alternative to the silage pit is big bale silage. The grass is cut and rolled into very large cylindrical bales which are inserted into equally large, sealable plastic bags. The bags may be stored outside. Small rodents can be a danger to them by piercing the plastic in search of a new home, and thereby letting air in. Big bale silage could seem an answer for the smallholder; there are snags, however - a front loading tractor is needed to move them, and, once opened, the contents must be used within a few days.

Machinery, implements.

The operations described in this chapter have included the use of a variety of tractor-operated machines and implements, including rotavators, furrow ploughs, underploughs, mole ploughs, disc, and chain harrows, rollers, spraying, and weed-wiping equipment, topping and mowing equipment, rowing, tedding, and baling machines for hay, big balers and multi-chop forage harvesters for silage.

To buy a tractor and a comprehensive range of this equipment, new, would turn any small farm into a loss making business. The purchase of a wide range of good secondhand machinery could do likewise for a smallholding. If you buy up a lot of cheap old machinery and implements, they will constantly need repairing. There is nothing worse than to have your hay ready for baling, hear an adverse weather forecast, and then find that your baler has broken down. You may consider starting with a reasonable secondhand tractor and trailer, and perhaps one or two implements that you can afford and consider essential. Amongst these, a roller can be important, since they are difficult things to move any distance, even if you can borrow one free of charge. For other jobs you may hire, or borrow equipment, or pay a neighbouring farmer or contractor to do the work for you. Starting off this way, you can gradually build up your range of implements whenever a bargain comes along. An alternative to this approach is to have neither tractor nor implements, except perhaps a roller for the reasons just given, hiring what you need when you need it, or calling in a contractor when there is a job to be done. The decision as to which approach to take will depend on economic considerations, whether you are mechanically minded and machine orientated or not, and on the availability of a contractor at a reasonable rate. One of us, with only twelve acres used for grazing sheep and beef cattle, and hay-making, found that he was only using his tractor during two weeks of the year - once in early spring for chain harrowing, rolling, and fertilizer spreading, and once later on to bring in the hay. He now has no tractor at all, finding it more economical to call in a contractor for the few essential jobs that crop up, and spreading fertilizer by hand. The choice is, however, very much a matter of personal preference - many smallholders like to feel independent through having their own machinery. Obviously the case becomes weighted in favour of your own machinery as the acreage increases and on holdings where some crops are grown every year. Whatever decision you take, think twice before compromising on a rotavator or similar machine designed for large domestic, or small market gardens - as soon as you took such a machine into a full sized field, you would regret it.

Strategic land use.

You should plan your land use to give you the best return, with avoidance of unnecessary expense and labour.

You may keep your labour to a minimum by arranging things so that stock such as poultry, which require frequent attention, are near your house and outbuildings. Your initial burst of enthusiasm will eventually decline and it may become a chore to walk, say, a few hundred yards twice a day to attend to half a dozen ducks, just because the site seemed attractive. Similarly, if you keep sheep, try to arrange things so that your ewes with their new lambs go out on a field close to the house where you can easily keep an eye on them.

We have already pointed out that sheep tend to deplete the phosphate reserves of the soil much more than cattle do. You can therefore keep your land in good heart, and get a better return from it, by grazing cattle and sheep in rotation. There is a further advantage if cattle and sheep graze fields in alternate years, for they are not susceptible to each others' parasitic diseases. If sheep or cattle are kept off a field for a year, their parasites in the soil die off. Using this system, cattle and sheep can go out each spring on to clean grazing. Cattle and sheep have each their own gut, and lung worms, and younger animals are more susceptible than older. Lambs are further subject to a parasitic worm called Nematodirus, which scarcely affects adult sheep. Land which has been used for conservation has had no stock on it in spring, so that the aftermaths offer relatively safe grazing, with no risk of heavy parasitic infection. From all this you should be forming an idea of the advantages of having a system of annual rotational grazing.

Grazing plans for cattle and sheep. If you have a holding on which all the fields are suitable for conservation, you may simply divide the land into three equal blocks, one for conservation, one for cattle, one for sheep, and have a three-year rota:

	Year 1	Year 2	Year 3
Block 1	cons	cattle	sheep
Block 2	cattle	sheep	cons
Block 3	sheep	cons	cattle

Cattle and sheep should be wormed at least a few hours before they go out on to the clean grazing. Later in the year, weaned calves and lambs can go on the aftermaths.

The scheme described above represents an ideal situation. Some compromise has often to be worked out if a larger acreage is given to sheep than to cattle, or vice-versa. The stock on the larger acreage would probably be given all the aftermaths.

On many holdings the conservation grounds are fixed because the other fields are unsuitable, having steep slopes or rocky outcrops. In such cases a two year rotation can be devised as follows:

	Year 1	Year 2
Block 1	cons	cons
Block 2	cattle	sheep
Block 3	sheep	cattle

Once again, some sort of compromise may have to be worked out if a larger acreage is to be given to either cattle, or sheep.

Grazing plans for cattle, or sheep alone. If all your land is suitable for conservation, you may be able to divide it into two equal blocks, on a two year rotation:

	Year 1	Year 2			Year 1	Year 2
block 1	cons	cattle	or	block 1	cons	sheep
block 2	cattle	cons		block 2	sheep	cons

Although this system does not provide clean grazing, it should provide safe grazing in normal years. Lambs or calves should be wormed several hours before they are put on the aftermaths. If you have a holding where all the land is suitable for conservation, this grazing plan can be quite profitable if you make hay. Nowadays many farmers make only silage; hay fetches a good price when it becomes scarce in winter.

If part of your holding is not suitable for conservation, you should try to arrange things so that *lambs spend the first half of the grazing season on fields which were not grazed by lambs in the first half of the previous grazing season, nor by undosed lambs in the second half of the previous grazing season.* If you keep cattle instead of sheep, *calves should not be put in a field before April 23rd. if that field was grazed by calves in the previous year.* This strategy will ensure that if your pasture is neither clean, nor safe, in normal years it will be acceptable, rather than potentially unsafe. If you feel unable to work to a planned grazing system, do not despair; try at least to adhere to the two last-mentioned rules for lambs and calves. Many farmers, in fact, keep sheep or cattle on the same land, year after year, successfully. They simply take greater risks and may have to spend more on worming.

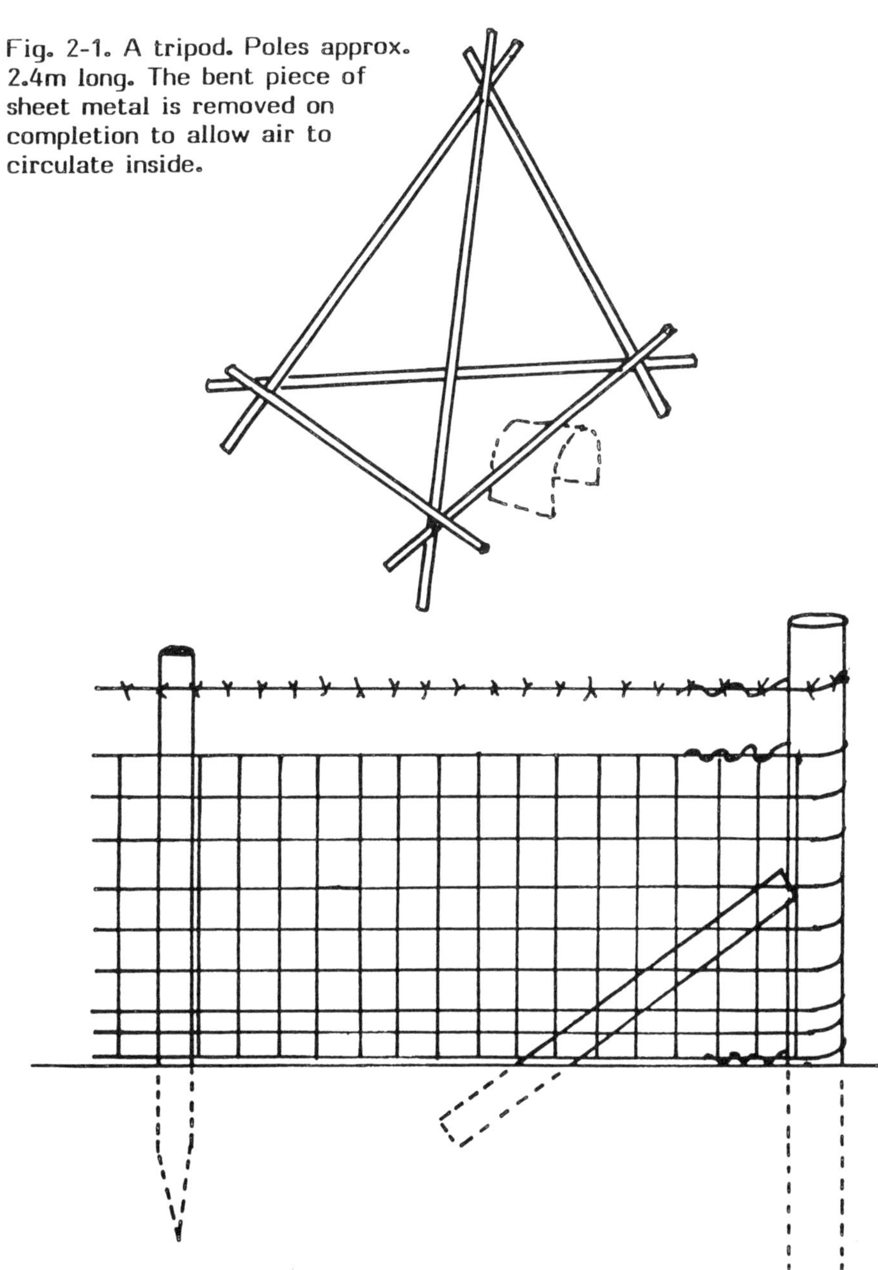

Fig. 2-1. A tripod. Poles approx. 2.4m long. The bent piece of sheet metal is removed on completion to allow air to circulate inside.

Fig. 2-2. Fencing using square-mesh galvanized netting.

Fencing.

Grazing efficiency can be increased by dividing each grazing block into at least three paddocks, and rotating the cattle or sheep round them. Two of the paddocks will be recovering whilst one is being grazed down. Some extra fencing may be required.

Good stock-proof fences can be made using strained square-mesh galvanized wire netting, sometimes called 'pig-wire,' and barbed wire. The netting comes in 50 m rolls, so it is best to set up straining posts at slightly less than 50 m intervals, and wherever the fence takes a change of direction. Straining posts should be of 2.1 m x 175 mm diameter treated timber, with angled struts of 1.8 m x 100 mm diameter. Posts and struts should be set well in the ground, and the earth compacted around them. The struts should enter the ground at an acute angle. Pointed stakes of 1.7 m x 75 mm square section, or equivalent half-round section, should be driven in the ground in a straight line at regular intervals between the straining posts. A roll of netting is laid out flat between the straining posts and about 60 cm of the bottom and top wires cut free from the rest of the netting. One of these loose ends is then bent round the bottom of a straining post, and wound round itself. The loop of wire is secured to the post by two or three stout staples. A strainer is used to strain this wire from the bottom of the other straining post. The free end of this wire is then looped and fixed to the bottom of the second straining post, as described above, using special fencing pliers as levers to keep up the tension whilst the staples are driven. The bottom wire of the fence is now fixed and strained. The wire strainer is detached. One end of the netting is lifted and similarly fixed to the post at a height corresponding to the width of the netting. The top wire is then strained to the correct height on the other post, and secured in place. Whilst the top wire is being fixed, it helps if someone holds the middle of the roll of netting upright; if no one is available, it may be secured temporarily to a stake or two by a loop of twine or lightly driven staples. The other end wires are then fixed to the posts, using staples, and the netting is stapled top, middle, and bottom to each stake. In the best class work, separate stout galvanized wires are strained top and bottom between the posts, and the netting is attached to these by metal clips. If you are keeping only sheep, a single strand of barbed wire, about 10 cm above the top of the netting may be adequate. For cattle, a second barbed-wire strand should be fixed about 10 cm above the first. Barbed wire should not be strained too tightly. Unstapled, strained barbed wire should never be cut, or suddenly released - it can whip through the air, wounding people or animals in its path. This kind of fencing is illustrated in fig. 2-2.

CHAPTER 3. CEREALS; FORAGE AND FODDER CROPS.

Some smallholders keep all their land down to permanent pasture. Whether you grow any of these crops will depend on the suitability of your land, and whether you have the implements for preparing and drilling a seed bed, and harvesting the crop in the case of cereals, or whether you can find someone to do some of these operations at an economic price. If your land needs re-seeding with permanent grass leys, you could think of doing it progressively over a few years, growing a cereal or forage crop undersown with grass, or prior to re-seeding with grass. If you undersow with expensive certified grass seed mixtures, you should go for a light cereal or forage crop, to ensure good survival of the grass seedlings. Some experts are of the opinion that it is better to sow certified grass seed mixtures after a cereal or forage crop has been harvested or grazed off. Some forage crops are suitable for ploughing in as green manure. This would be a useful procedure, prior to re-seeding, on an organic farm.

Cereals.

Cereals should not be sown in small plots, for the local birds will decimate the seedlings. When a whole field of cereals are sown, the birds take their share, but leave sufficient young shoots to provide a crop. It is unwise, therefore, to think of growing a small plot of cereals to obtain grain for home baking. The majority of smallholdings in Britain do not occur on the best arable land. It is unlikely, therefore, that you will be able to grow cereals as a cash crop. If your land is suitable, you might think of growing a few acres of oats, or more likely barley, to enable you to cut down on winter feed costs, and to provide straw for feed or bedding. If you are unsure about the suitability of your land, try asking some of the older local farmers who may recall what has been grown there in the past. You should note also that your grain may need drying so that it will store without going mouldy. We know one farmer in Wales who grows a light crop of barley, cuts it green, and dries it like hay as a high protein winter feed. His sloping land faces west so that it is well-ventilated and warmed by the summer sun until quite late in the evening. We mention this procedure in passing; we would not recommend it to the inexperienced, for cereals are much more difficult to dry than grass. Cereals can also be planted as forage crops; forage rye sown in September or October can provide a good 'early bite' in spring, but this is more often used by dairy farmers as a way of keeping up milk production.

Forage and fodder crops.

Forage crops can be grown to provide good quality feed in summer, and to provide feed during late autumn and winter when any remaining grass has little food value. Many of these crops have different fertilizer requirements; if you are not an organic farmer, you should check the requirements when buying the seeds. Some root crops can taint cow's milk; if you have a house cow you should look through the seed catalogues for varieties which are not supposed to taint milk. Some forage and fodder crops are listed below:

Forage peas can be sown on light, well-drained soils. They are normally grown for silage, but they can be control grazed by sheep or cattle using an electric fence. They are suitable for undersowing with grass. Forage peas leave a fair amount of residual nitrogen. They can be beneficial, therefore, on organic farms, if the soil is suitable for them.

Mangolds and **fodder beet** can be sown in the spring to produce large yields of dry matter for cattle in the winter. The tops can be grazed off in October and November, and the crop lifted and stored, to be fed, chipped, to cattle during the winter. The tops of fodder beet have a high sugar content and they can be converted easily into silage in a simple clamp.

Swedes can be sown in late spring to provide fodder throughout the winter.

Forage rape is a fairly fast maturing crop. There are varieties which can be sown in spring to mature in July and August, and others to be sown in July and August to mature in October and November. A field of rape can be useful for fattening lambs - your own, or ones bought in from a mountain farm where they cannot be finished. Summer sown rape can extend your grazing into early winter, when the food value of grass is poor.

Turnips can be sown in spring, for summer grazing, or in late spring and early summer for autumn and early winter. Stubble turnips don't have to be lifted and they are very good for sheep. There are other grazing turnips, such as tyfon, obtained by crossing Chinese cabbage with stubble turnip, which can be matured quickly in the growing season, or sown in early September for November grazing.

Field cabbages can yield up to forty tonnes per acre. There are early and late varieties. Both need to be planted in spring so they are not so fast growing as rape or turnips. They are best cut and carted; grazing them is less economical. They can provide fodder from July to February, if you sow a range of varieties.

Kale. Different varieties can be sown in June and July to provide forage greens from September until late February. As with other forage crops, controlled grazing, using an electric fence, prevents wastage.

Forage mixtures. Some seedsmen provide 'forage mixtures' of seeds; they are specially suitable for the smallholder who may wish to plant only one, or at most two fields of forage crops. There are **'autumn keep'** mixtures of rape and stubble turnip which may be planted any time between mid-April and August to provide forage from midsummer to the end of the year, or for ploughing in as green manure in the autumn. There are also **'winter green'** mixtures of kales and rape for sowing in June or early July to provide forage from September to early February.

CHAPTER 4. SHEEP.

Breeds of sheep.

There are many different types of sheep in Britain. Each breed is best suited to a particular set of environmental conditions (climatic, topological, etc.), and each has its own special characteristics of behaviour, carcass quality, and type of fleece. From a commercial point of view the carcass is much more important than the fleece, since the return on the latter is relatively small. Broadly speaking, the various breeds can be classified as mountain, hill, and lowland types. Mountain sheep should only be kept on mountains, for they become escapists when kept in small fields at lower altitudes. Similarly, lowland sheep should not be kept on the hills because they are not hardy enough.

In choosing which breed, or breeds, to keep it is important first to ensure that they are suitable for your holding, and second, that they will be a saleable commodity at local markets. A few visits to local marts can be useful to settle the last point. A few conversations with local farmers may help you decide which of the popular breeds in your area will be suited to your holding. Try to avoid forming a preference for a breed of sheep that is not commonly kept in your area, for, even though it may be adapted to the conditions on your land, it is likely to be less profitable. For example, if you were to keep a small flock of Swaledale or Wiltshire sheep in Wales, or Welsh Llanwenog sheep in Cumbria, although you may sell the lambs for meat as profitably as those of a local breed, you would not obtain the right prices at local marts if you needed to sell some ewes, either in a financial crisis, or because you had bred replacements for them, simply because the local people would not know them.

If you wish your sheep flock to be profitable, it is also important to avoid the temptation of keeping unusual breeds such as the Soay or Jacob's. Fascinating as these breeds are, they should be left to those who can afford to keep them (and who thereby do a useful service by maintaining an extended sheep gene-pool). Although these breeds are hardy, they can be wild and difficult to handle. One of us started off with a few Soay sheep, after reading somewhere that they were 'the sheep of the future.' Five minutes after they had been released into their field, they had disappeared, reappearing later in an adjoining field. It transpired that they had pulled a fencing stake out of the ground by inserting their horns in the fencing wire, and lifting. They proved almost impossible to round up, feared neither people nor dogs, and readily jumped fences. Since that experience, he keeps only popular local breeds.

Pedigree, pure-bred, or cross-bred sheep.

With a small acreage it is tempting to try to increase the profit by keeping a pedigree herd. There are, however, problems and risks, especially for the inexperienced farmer. Pedigree breeders have built up a reputation for themselves over the years, and farmers have confidence in their products; the same farmers may be reluctant to purchase equally good sheep from a newcomer to the market. Pedigree breeders also have acquired the art of making their animals look good, something which cannot be picked up overnight. Your pedigree flock will cost a great deal of money, and this means that any losses resulting from inexperience may, at best, make a potentially high-profit flock more troublesome and no more profitable than an ordinary commercial flock. Even with ordinary commercial breeding, losses of five per cent per year are normal. It is, however, possible to compromise by keeping a mainly non-pedigree flock with a good pedigree ram and two or three pedigree ewes of the same breed. In this way a start towards pedigree breeding can be made without a great capital risk. The pedigree ram could be an elderly one, and therefore not too expensive. There is, however, a snag, for most breed societies will only admit a breeder with more than a specified minimum number of pedigree sheep. This means that for most breeds you would have to sell your progeny as 'pure-bred, non-registered sheep, from pedigree stock.' They will not, therefore, fetch as high a price as registered sheep, but this is nonetheless a means of making a start at pedigree breeding, with minimum capital risk.

If, instead, you decide to go for pure-breeding, you will be selling your ram lambs and some of your ewe lambs as meat, but the best of your ewe lambs will either be sold to become breeding ewes, or kept as your own flock replacements. In the latter case, you will be selling off your older ewes as they are replaced. If you rear your own replacements, you will, of course, have to avoid mating them with their father, which means keeping at least two rams, or replacing your ram every year.

Many of our mountain and hill breeds of sheep, when pure-bred, produce lambs with good quality meat but only a low carcass weight. Some farmers who keep these breeds cross the ewes with a lowland breed of meat ram (such as the Suffolk, Dorset Down, Hampshire, Shropshire, etc.) to give a larger, faster growing lamb which will grade with a higher carcass weight. At moderately low altitudes the cross-bred ewe lambs can be kept on as breeding ewes, but none of the ewe lambs from these cross-bred ewes should be kept as they may lack hardiness. If you keep any ewe lambs as replacements, they must first be mated with a smaller hill breed of ram. If they are mated, for their first lamb, with a large breed of ram, they are likely to have serious lambing difficulties. Sometimes cross-breeding is carried out with a view to producing breeding ewes of special quality. For example, Blue-faced Leicester rams are crossed with hill breeds, and the resulting ewes, known as 'mules,' make excellent breeding stock and fetch a high price from farmers at low altitudes. The problem with this particular cross is that the Blue-faced Leicester breed is quite delicate and the rams need careful attention in hill country.

Buying sheep.

Sheep may be bought privately, at a farm sale, or at a livestock mart. Breeding ewes are not often advertised for sale privately, but breeding rams and ram lambs are often advertised in late summer and early autumn. Farm sales may be reduction sales, or dispersal sales. A reduction sale generally takes place when a farmer is selling off part of his flock (not always the best part) as a result of a change in his farming policy. A dispersal sale involves the sale of the entire stock on a farm, either because the farm is being sold, or the farmer is retiring and letting out his land to other farmers (tack farming). The advantage of buying privately, or at farm sales, is that there is less risk of disease in the animals, compared with those from a market. The advantage of buying at auction mart is that there is a wide choice, and prices are often lower than at farm sales.

Wherever you buy, if you are doing so for the first time, it is a good idea to ask an experienced local farmer to go along as advisor. Sometimes a retired farmer, who is looking for something with which to occupy himself, will be happy to help in this way, and, quite often, he will only want his petrol money or a couple of pints of beer for the service. Beginners often make the mistake of looking only at the general condition of the sheep, and seeing that their faces are attractive, when in fact, they should be checking their teeth, udders, and feet.

Rams and breeding ewes are sold at all ages. When buying a ram, look for a good body conformation - it should be like a shoe box on its side with a leg at each corner. Avoid any rams with unusually wide heads for their breed. If you are inexperienced, it is advisable to buy a ram that already has some experience himself, rather than start with a ram lamb. If cost is a consideration, go for an older ram with a good conformation, rather than a younger one of inferior quality. Ewes are sold as ewe lambs, yearlings (shearlings, or hoggets), two year olds, etc., up to 'full mouthed' and 'brokers.' Yearlings have two broad central teeth, and milk teeth either side;

two year olds have four broad teeth; older sheep six, and then eight broad teeth (the full-mouthed condition); brokers have some teeth broken or missing. The inexperienced are advised not to start with ewe lambs, nor with shearlings which have not lambed as ewe lambs, for they are more likely to have lambing difficulties. The decision whether to buy younger ewes (say, two, or three year olds), full-mouthed, or brokers, depends partly on how much you wish to spend, and partly on your sheep-farming policy. In any given breed, younger ewes cost more to buy, but less to feed in winter, whereas brokers are much cheaper, but cost more to feed. There are those who maintain that the most profitable method of sheep farming away from mountains is to buy in brokers every year. Some farmers buy in a fresh set of brokers every autumn, and sell them with their lambs the following spring. They can then use their land in late spring and summer for other purposes (conservation, cattle, etc.), but this method may disqualify them from subsidies. Perhaps the best advice for the beginner is to start with good full-mouthed ewes. They are a little more expensive than brokers, but they are fully experienced mothers and may still have a few years' good service left in them.

As soon as you get your sheep home, put your chosen mark on them, if possible using a colour different from that used by your closest neighbours. In addition to colour marking, you may also use a patterned ear punch as a precaution against rustlers, but this may be too expensive for a small flock.

Rounding up.

On a mountain, or large hill farm, trained sheep dogs are indispensable. They are not, however, essential on a small acreage with a flock of, say, up to forty ewes. A newly-purchased flock will be difficult to round up in its first season, but, once you have fed them a few times in winter, and they have become used to your sight and voice, they will generally follow when you call them, especially if you carry a bucket containing a little concentrate feed. From then on, rounding up should present no problem, except perhaps on windy days, when they can be difficult to manage.

Penning in; sheep-handling complexes.

Sheep are reluctant to go into any enclosure from which they cannot see a way out. Trying to lead, or drive them through a doorway into an enclosed barn or shed generally proves difficult. It is much easier to pen them in if you have two gates (or sets of hurdles) a short distance apart on a short fenced lane leading into a fenced yard or small paddock. The sheep will readily run past the open gates into the yard or paddock, and the first gate is then closed. The sheep are then driven back to this gate, and the second gate is closed, trapping the sheep in the pen between the gates. It is important to use normal farm gates or fencing-type hurdles, and not solid ones, as sheep run more readily to a gate or hurdle that they can see through. For a small flock, of up to about forty ewes, this type of pen can be adapted for sheep-handling purposes, simply by the addition of a couple of hurdles to divide the space within in various ways, and a portable foot bath (fig. 4-1). For a larger flock, a more elaborate sheep-handling complex is a labour-saving advantage. The basic elements of such a complex are:

1. A **catching pen** which can be used for trapping and holding the sheep. This pen should be planned with gates or hurdles which can be used to reduce the space progressively, so that, as sheep are moved out, the remainder can be kept tightly penned together. Penned sheep can panic if given too much space in which to move around. Large complexes sometimes incorporate a circular catching and forcing pen.
2. A **sheep-dip bath** which can be shut off, or boarded over, when not in use. A concrete draining pen must be arranged at the exit side of the dip, where surplus dip solution can drip from the sheep and run back into the bath.
3. A **footbath**, often with a remote-operated guillotene gate near the exit, arranged so that a single operator can push the sheep in at one end, and control their exit by pulling on a rope passing over pulleys to the gate.
4. A **drafting race** opening into two holding pens. The gate at the end of the race can be flipped to one side or the other, permitting easy and rapid separation of the sheep. This is useful for separating lambs from their mothers, or sheep destined for the market. A removable footbath is often incorporated in the drafting race.
5. A **working race:** Since the drafting race has to be narrow, sometimes an additional working race is incorporated alongside the drafting race. This is made wide enough to permit a worker to handle the sheep, one by one, as they pass along.
6. **Decoy pens:** Sometimes one or more small pens are strategically placed, each holding a decoy sheep, to encourage the other sheep to move through the system.

A sheep-handling complex suitable for a medium sized flock is shown in fig 4-2. Naturally, there are many variations and each complex has to suit the nature of the site chosen and the size of the flock. All the fences of the pens and races must be robustly constructed of stout timber or galvanized tubular steel. Wire fences are quite unsuitable for this purpose. This does not necessarily mean that the fencing will be expensive; rustic timber can be

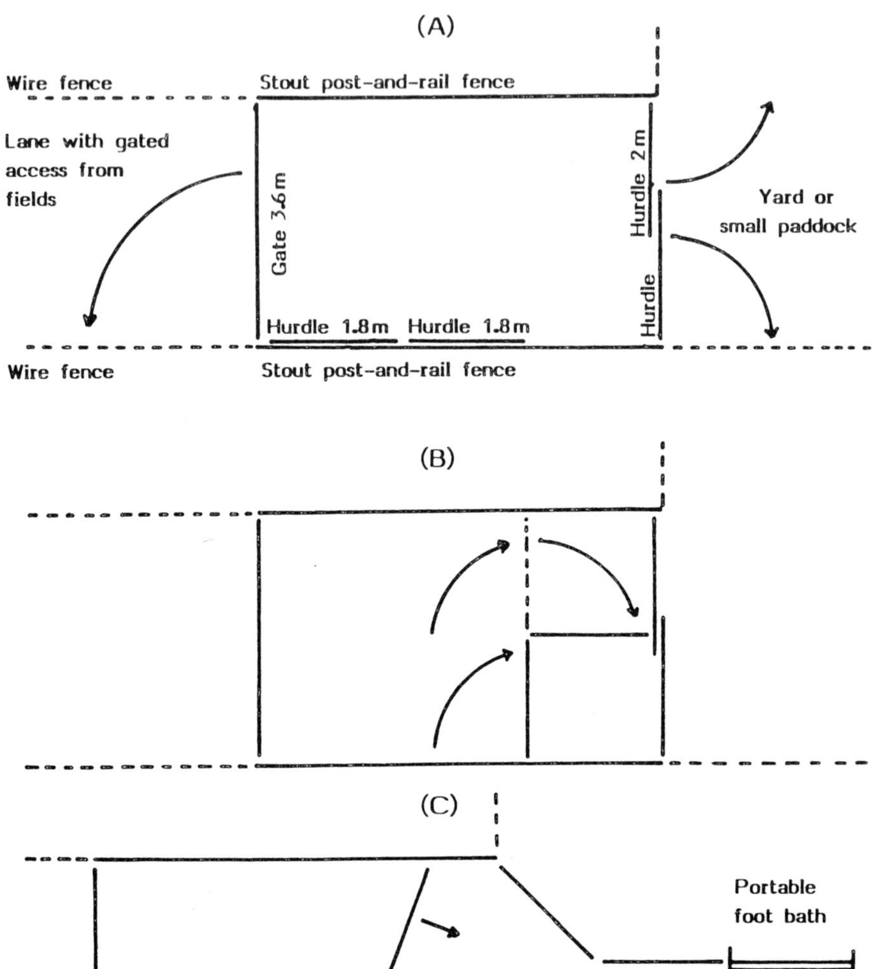

Fig. 4-1. A simple sheep pen. (A) Diagram of basic lay-out.
(B) Ways of using hurdles to separate off some of the sheep.
(C) Pen with temporary race and portable foot bath added.
String or rope ties are used for temporary attachments.

used on condition that the rail-to-post attachments are securely made. If you wish to use treated sawn timber, the posts should be 1.5m x 100mm x 75mm, set a good 45mm in the ground, spaced at 1.2m intervals, with rails made of 75mm x 38mm timber. If you think you can claim grant for a sheep complex, you should first check that your specifications will be approved. Non-slip concrete floors are useful throughout a complex, but they are most important around the dip bath, in the draining pens, and in the races. If you are making a complex on sloping ground, arrange things if possible so that the sheep go uphill as they move through the system. It is always easier to induce them to go uphill, for this is their natural escape reaction in the wild.

Sheep handling.

When catching sheep, aim to grab them in a sort of bear-hug; avoid grabbing them by the wool, for this can cause a superficial bruise just below the skin and devalue the carcass of any animals taken for slaughter soon afterwards. Sheep can be 'turned up' and held immobilized in either of the following ways:

1. Stand on the sheep's left side and pass your left hand beneath, behind the sheep's left front leg, and grip the middle of the sheep's right leg; then lift, at the same time grabbing and lifting the middle of the sheep's right hind leg with your right hand. When done properly, the sheep ends up sitting on its bottom with all four legs sticking out. You can then hold the sheep steady between your knees, and have both hands free for performing whatever operation is required - foot trimming, etc.

2. Stand on the sheep's left side, gripping under the tail with your right hand and pushing the head away from yourself and round towards the tail; this causes the sheep to roll over your leg and sit on its bottom. You can then hold the sheep between your knees as described above. This method is the least strenuous, but it takes a little while to acquire the knack.

A few farmers who keep the heavier lowland breeds, handle their sheep by lying them on the ground - they tend to struggle in this position and have to be kept pinned down by the shoulder.

There are patent sheep-handling crates on the market which will do the job for you. The sheep has to be induced into the crate, and then a lever is operated to tilt the sheep into a helpless position. This kind of apparatus is expensive, but it could be a great help for those who are getting on in years, or who suffer from back trouble. A much cheaper alternative is to use an ordinary deck

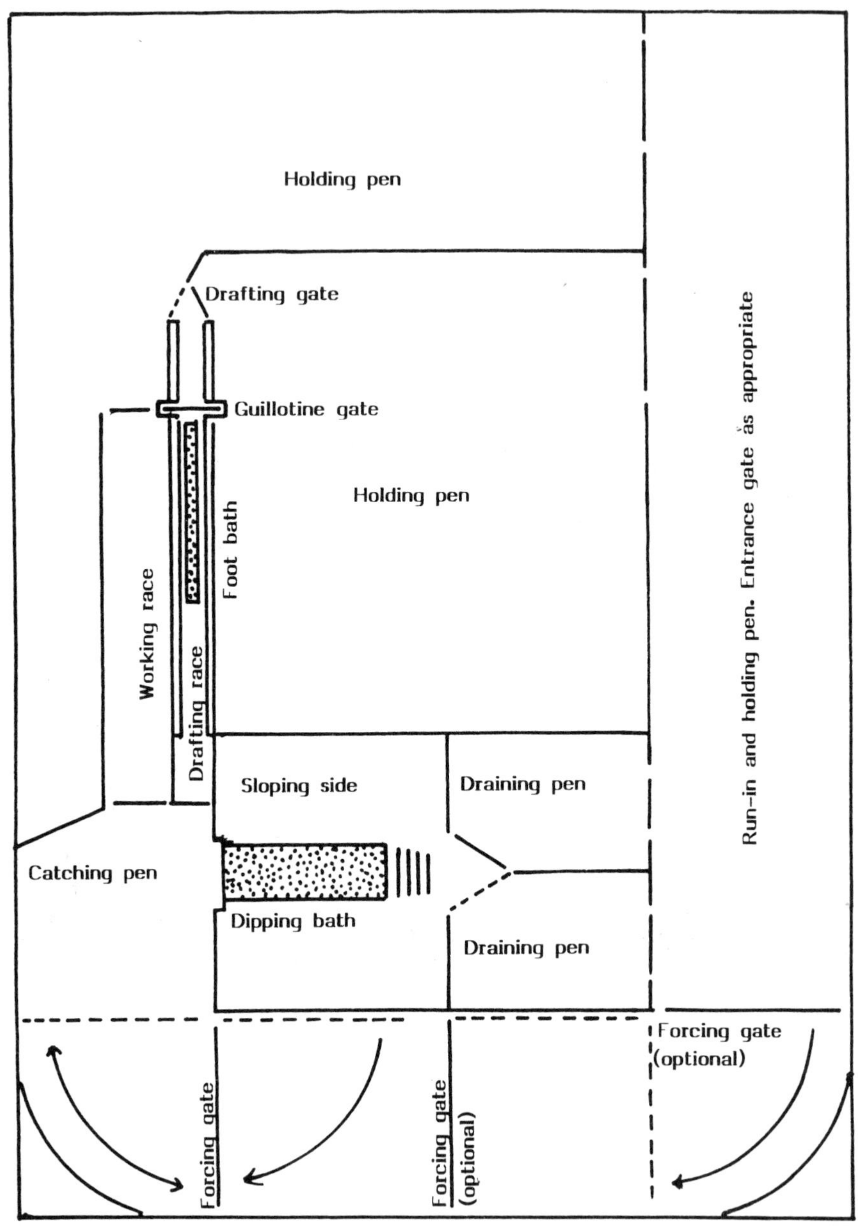

Fig. 4-2. Sheep handling pen suitable for a medium sized flock. Scale approx. one hundredth, but size depends on number of sheep. For a large flock a circular forcing and catching pen is an advantage, as well as larger draining pens.

chair - set up at the right height, it will hold a sheep in a helpless position, but there still remains the hard work of tilting the sheep into place.

Keeping an eye on the flock; routine operations.

Sheep should be inspected daily on their pasture. Failure to do so can result in losses from sheep which get caught up in brambles, fences, or hedgerows, or those which have rolled on their backs in a small hollow and are unable to right themselves. When you see any sheep that look sick or are limping badly, it is easiest to bring the entire flock into the sheep pen, and separate them off. Trying to catch one or two sheep in a field is much more difficult. Your sheep will also have to be brought in periodically for foot trimming and protection against foot rot, vaccination against various diseases, drenching (worming), dipping, shearing, etc.

Foot trimming.

Just like our nails, sheeps' hooves grow continuously. In nature they are worn down by the sheep scrambling over rocky mountain sides. On pasture, they have to be trimmed periodically. You can do this with a paring knife, or special foot shears. It is best to have this operation demonstrated to you by someone with experience, as cautious beginners trim off too little, the others too much. During this operation, watch out for stones stuck in the cleft between the feet, and for signs of foot rot. The latter is easily detected by its foul smell. All the affected parts must be carefully pared, or cut away, stopping only when there are signs of bleeding. If the flock is not about to be put through the foot bath, you should treat the infected foot with an aerosol spray. There are several kinds of spray on the market; the one we prefer contains oxytetracycline and gentian violet. You may need to repeat this process a few days later. Instead of using an aerosol spray, some people stand the infected foot in a small container of diluted formalin. This practice is dangerous - if the sheep suddenly kicks out, you can end up with formalin in your eye, as happened to one of us.

The foot bath.

Your entire flock should periodically be driven through the foot bath as a measure intended to reduce the incidence of foot infections. The most commonly used chemical is formalin, and this must be diluted according to the manufacturer's instructions. We have heard of people putting the undiluted liquid in the foot bath, with dire consequencies to their sheep. If the sheeps' feet are muddy, it is best to run them through a shallow container of water before they enter the foot bath. It is particularly important to use the foot bath in autumn, because, with the onset of wetter weather there is greater risk of foot infections spreading.

Ready-made foot baths may be purchased, in metal or glass-fibre. If you purchase one, avoid any that have a flat floor; those with ribbed, or non-slip convex floors help to splay out the sheep's feet so that the liquid penetrates the cleft. Remember that these commercial foot baths generally require the construction of a sloping-sided race to house them. As an alternative, you can make an integral portable foot bath from waterproof plywood, using any of the glues used in boat building. Construction details are given in fig. 4-3. The one we made is now entering its fourth year of service, with no signs of deterioration.

Vaccination.

Routine vaccination can be carried out to protect your sheep from a variety of diseases, especially the clostridial ones, and also against foot rot. There are multiple vaccines on the market which protect against a number of disease organisms with a single injection. You should consult your local vet. about which diseases are prevalent in your area, for there is no point in vaccinating against those that do not commonly occur there. In theory, a vaccination programme against foot rot should make it possible to eliminate the disease from a holding, for the causative organisms cannot survive more than two weeks in the soil. This means that animals free from infection only have to be kept in fields that have been sheep-free more than two weeks, to have no risk of picking up the disease. In practice, the disease is more difficult to eradicate than one would think. Further, the manufacturers of the foot rot vaccines point out that it is still necessary to use the foot bath to protect from other foot diseases and conditions. On balance, however, the incidence of foot rot is often markedly reduced after a complete vaccination programme. - For injecting sheep, use a short, fine needle and always check where each injection should be given, and whether it should be subcutaneous, intramuscular, etc. When vaccinating a number of sheep, a gun-type hypodermic is an advantage; permanent ones are fairly expensive, but there are plastic throw-away kinds. As you inject each sheep, mark it with a coloured crayon to avoid any risk of doing it twice.

Drenching, (dosing, worming).

If your land is well-drained you will only need to dose your sheep against roundworms and tapeworms. If, however, there are permanent wet areas on your land, you will also need to dose against

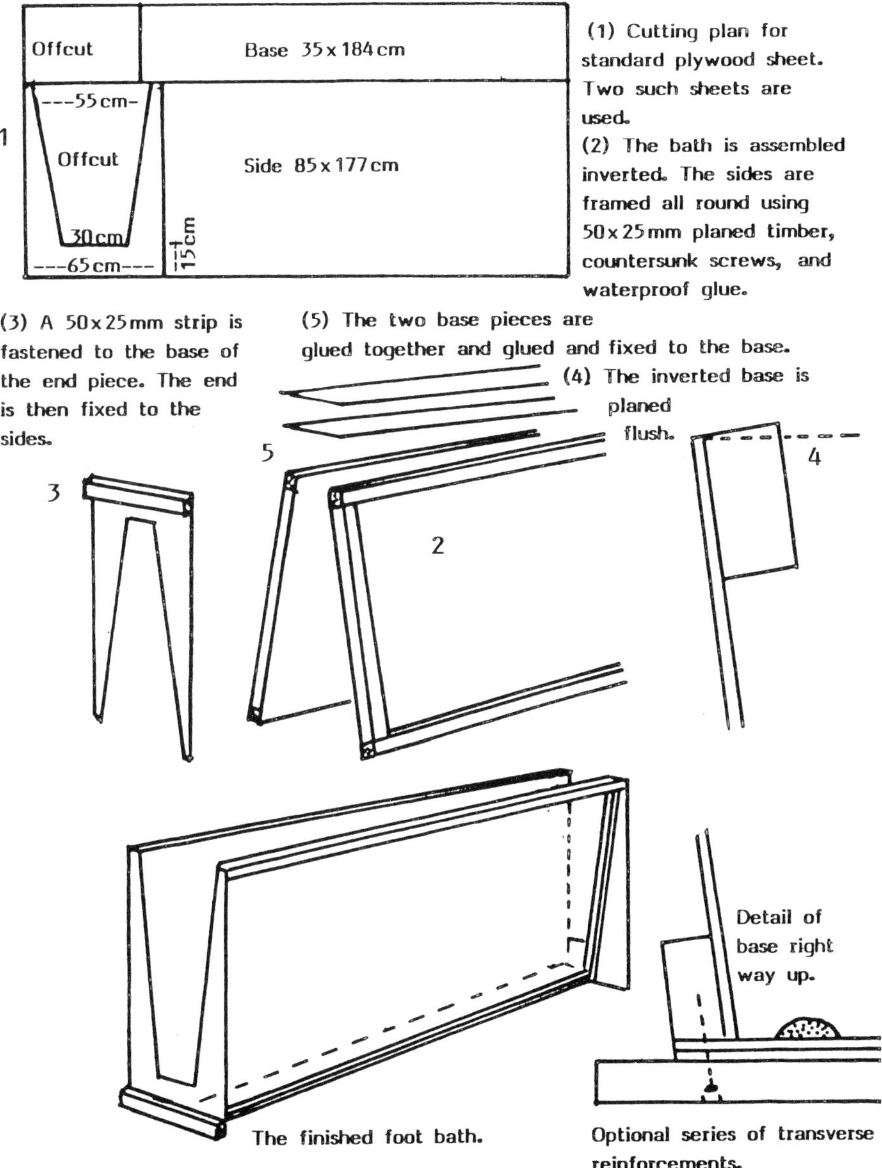

Fig. 4-3. Portable integral foot bath made from 50 x 25 mm planed timber and two sheets of exterior ply, 6mm thick. Strength may be increased by using 9mm or 12mm ply, and the length may be extended to 2.4m, using three sheets of ply. All contacting surfaces to be glued using waterproof glue suitable for boat building. Half-round timber strips should be glued inside the base to help splay out the sheeps' feet.

Allow the sheep to right themselves, with their heads clear of the liquid, and keep them thus for a full minute before letting them climb out into the draining pen. This satisfies the present legal requirement, and it ensures that your dipping will be an effective preventative remedy. Do not dip sheep with wet fleeces, for the solution will not penetrate properly. If it rains, wait until their fleeces have dried off.

Shearing.

There will always be a few enthusiasts who will insist on learning to shear their sheep with hand shearers, and good luck to them. Machine operated shearers do the job much quicker. It takes quite some time to acquire the necessary combination of skill and speed for shearing, and new electric shearing equipment is fairly expensive. For a small to medium-sized flock, the cost of shearing by a professional team may be little more than the interest you would pay if you borrowed money to buy the equipment new, and there is a bonus in work saved. We therefore recommend you to call in a professional team to do the job, at least in your first few years. The sheep should be penned in such a way that they can easily be caught and brought to the shearing area. The latter should be provided with a clean floor - a large tarpaulin, or similar, is ideal for this. Sheep cannot be shorn if their fleeces are wet, so, if possible, they should be penned and shorn under cover. As soon as the shearers remove a fleece, it should be taken to a clean floor or large table and laid shorn-side down. Any dirty bits of wool or debris should be removed, the sides folded in towards the centre line, the fleece rolled up, and the roll secured by a length of twisted wool. Cleanliness is the essential in this operation - fleeces with dirt or bits of hay or straw may be devalued. In Britain, if you own more than three sheep, you are legally required to sell your wool to the Wool Marketing Board*. You should register with them as early as possible, to give them time to process your registration and inform you, via their agents, of the date(s) when you may take your fleeces to a collection point.

Tupping (mating).

The larger breeds of ram can work more than forty ewes, and some of the wilder mountain rams a good twice as many. If, however, you wish all your lambs to be born within a reasonable period of time, you should not exceed forty ewes to one ram. Ram lambs should only be given about half the number of ewes. This means that if you only have a small flock, one ram may be sufficient, in which

*Oak Mills, Clayton, Bradford, W. Yorks. BD 14 6 JD.

case it is important to be sure of his fertility. As a safeguard, in your first year, you could have your ewes scanned for pregnancy, or you could perhaps borrow a neighbour's ram at the end of the tupping season, in case your own ram has not been a 'good worker.'

Gestation takes five months, and the oestrus cycle three weeks. You should therefore put your ram or rams with the ewes around five months and three weeks before you intend to lamb. The ewes should be in good condition, for tupping. If you wish to lamb early, note that many hill and mountain breeds will not come into oestrus before autumn, even in the presence of a ram. If you wish to synchronize your lambing as much as possible, you should first run your flock with a vasectomised teaser ram. This will ensure that your ewes will be cycling synchronously when you later admit the working ram or rams, just five months before you intend to lamb. The working rams may be raddled - either by rubbing their chests with a coloured powder mixed with oil, or by fitting a raddle harness on them. Raddling ensures that each time they mount a ewe they will leave a coloured mark on her rump. If you are breeding replacement ewes, each ram can be raddled with a different colour and the ewes can be ear tagged to avoid any danger of mating the next generation of ewes with their own father, if they, also, are tagged. Once a raddle mark appears on a ewe, do not assume that she has conceived; she may need mounting again at the next oestrus, or later. You can always change the raddle colours so that you can note the ewes that have been mounted a second time and are going to lamb later. Beware of making a pet of a ram - it is best if he respects you, for, once he loses his wariness with you, he will butt you at most unexpected moments, and could injure you.

Feeding; housing.

For much of the year your sheep will live on pasture alone. It is, however, advisable to provide them with mineral licks throughout the grazing period. This is very important in those parts of the country where the soil is deficient in magnesium. Grass staggers (hypomagnesaemia) results from sheep eating grass deficient in magnesium. The period of greatest risk is when the grass is growing rapidly, and the risk is therefore increased with a high nitrogen fertilizer regime. Death from grass staggers can be quite sudden, but a collapsed animal can be treated by a magnesium-rich injection, if spotted in time. Mineral licks may be liquid, containing a molasses base, powder, or rock-hard solid. If you use a solid lick, hang it from a projecting piece of wood nailed to the top of a fence or gate post. Your sheep will then be able to lick it, but not bite on it; this protects their teeth.

At some time during the autumn or winter you will have to start feeding concentrates and hay. If your sheep are outdoors you should start feeding hay when there seems to be no more useful grazing for them; if they are housed, they will need hay all the time. You should start feeding concentrates at least five or six weeks before you expect to lamb, starting with four ounces per head per day, increasing gradually to about one pound per day the week before lambing. This is only a rough guide, for the precise rations will vary with the breed, the condition of the flock at the onset of winter, and the age of the sheep. Brokers require more feed than younger ewes, and those having twins require more than ewes having single lambs. It is important to continue feeding ewes after lambing, gradually reducing the quantity as good spring grass becomes available. It may be a good idea to discuss rations with those of your neighbours who keep the same breeds, both profitably and in good condition. As a check, once you have decided on rations, add up how much concentrate you intend to give each ewe during the weeks before and after lambing, and work out how much this will cost you per ewe. If you find that the cost per ewe in any way approaches the expected value of her lamb(s), then your farming is not going to be profitable, and you have over-estimated the rations. If, on the other hand, you don't feed enough, your ewes will go out of condition, and your farming will again not be profitable. Your ewes will also run the risk of twin-lamb disease (pregnancy toxaemia). - When you are feeding manufactured concentrates there is little need for mineral licks. The concentrates contain the correct proportions of minerals. Ensure that the feeds you use are for sheep - those manufactured for other animals have a different mineral content - for example, cattle concentrates contain copper in a proportion that could be harmful to sheep. If you are feeding grain or sugar beet pulp, instead of factory-made feeds, you should still provide mineral licks.

Outside, concentrates can simply be poured in a long, thin row along wooden or metal feeding troughs (fig. 4-4). You should have enough of these troughs to allow all your sheep to feed at once, otherwise the timid ones will lose out. Indoors, concentrates can be fed in similar troughs, but they must be kept clean. If you have an indoor walk-through hay box system, this may be used also for feeding concentrates, daily, when they have cleared up the hay. Hay boxes may be constructed following the diagrams in fig. 4-4. They may be arranged, with hurdles, to form separate indoor pens. As the hay boxes can be walked through, you can distribute the feed without having to push through the penned sheep. These hay boxes are better than inclined hay racks - the sheep have less risk of getting seeds in their eyes, and less hay is wasted through being pulled on the ground and trampled on. For use outside, these hay

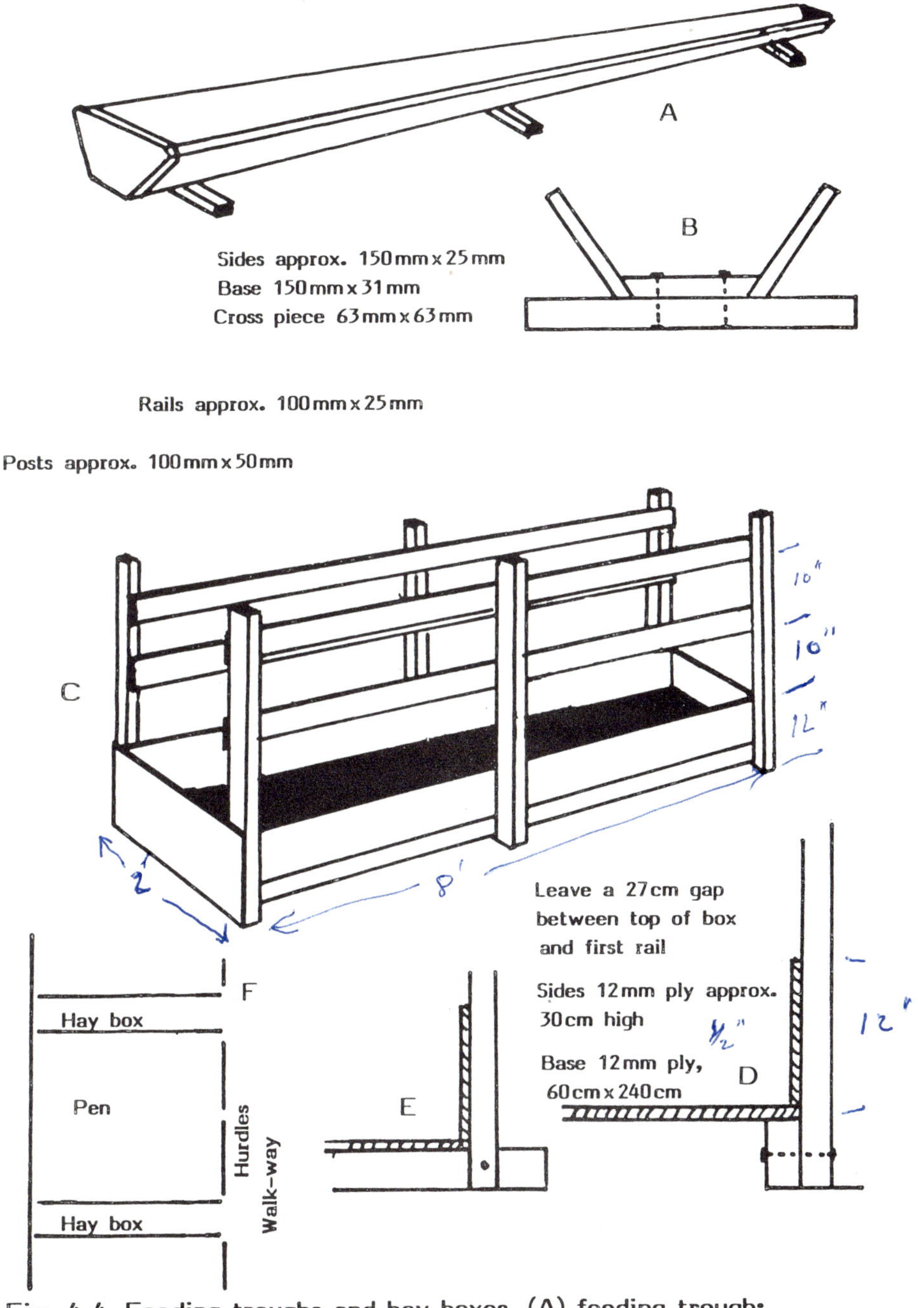

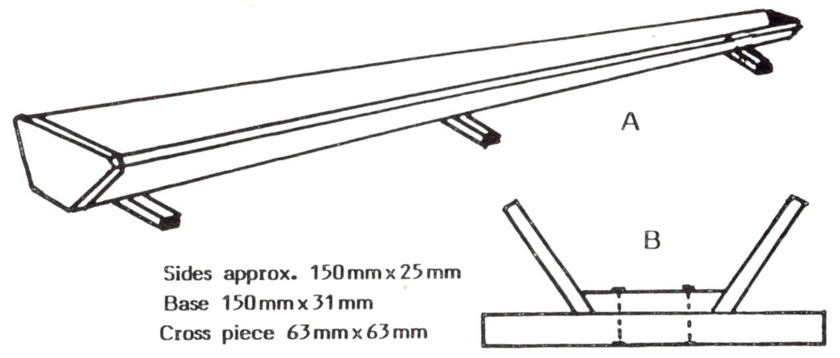

Fig. 4-4. Feeding troughs and hay boxes. (A) feeding trough; (B) cross-section. (C) Hay box; (D) section through part of base; (E) section showing alternative construction. Note – edges of feed box and lower rail should be rounded off. (F) Pens for in-wintering.

boxes should be fitted with a removable, overlapping waterproof cover, such as a corrugated sheet.

Many sheep farmers keep their flocks outdoors throughout the year, moving them to fields close to their house at lambing time. A few, mostly lowland farmers, in-winter their flock. In-wintering is probably economical for the larger, and more remunerative breeds. Apart from the cost of housing, there is the work of clearing out the manure and spreading it back on the land. Slatted floors can be provided to make mucking out a once-a-year operation, but they further increase the cost of housing. A reasonable compromise is to have your sheep in only at lambing time. Using an existing building, temporary pens can be set up, using hay boxes and hurdles as described above. Straw, or wood shavings should be spread on the floor, and added as required. Drinking water should be available at all times. This will make lambing easier for you, and it will leave your barn or shed free for other purposes for the rest of the year.

If you have no suitable building, and propose to erect one, it need not be too expensive. The walls may be of concrete blocks 1.5m or 1.8m high, followed by space boarding, and a roof of corrugated sheets. If an apex roof is constructed, allow a gap near the ridge to let out the warmer, humid air as it rises up. The essentials of such a building are a relatively large volume of air, good ventilation, and no draughts around the animals. An alternative to concrete blocks is to use shuttering plywood fixed to posts, with space boarding above it. Cheaper still, you could use old colliery conveyor belting nailed horizontally to the posts right round the outside of the building, up to the height of the space boarding. This will be no more unsightly than many of the black-painted corrugated outbuildings that one sees around many farms. If you use material like conveyor belting, it would be advisable to fasten two or three timber rails to the inside of the posts at a height that would prevent sheep from butting the outer skin. In all timber-framed buildings, the timber posts must be firmly set in the ground, or bolted to long steel ties which are set in concrete. Similarly, the roof timbers must be securely tied to the uprights. These suggestions are made as a guide only - if you think you may be able to claim grant on a building, you should first ensure that your specifications meet approval for that purpose.

Lambing.

Good management at lambing time contributes more to a successful sheep enterprise than at any other time of the year. If you have no experience of lambing it is a good idea to attend lambing classes - the Agricultural Training Board provide courses at various locations.

Whether your ewes belong to a prolific breed or not, what matters is that you obtain the best possible lambing percentage; this depends on the number of lambs successfully reared per hundred ewes, and not on the number born.

Early lambing is important for pedigree herds because larger ewe and ram lambs will fetch the best prices when sold as breeding stock the following autumn. When lambs are being reared largely for the meat trade, early lambing produces lambs which will grade in the spring when prices are high, but the cost of feeding the ewes will be greater; late lambing, when the ewes can soon be feeding on new spring grass, gives lambs that will grade when prices are lower. At high altitudes lambs often fail to grade for meat, and they have to be sold for finishing, as stores, to farmers at lower altitudes. Lambs kept on to grade during the winter months often fetch a good price; a field of rape or other fodder is best for them; otherwise they have to be given hay and (expensive) concentrates. If you buy lambs for finishing, ask what they have been feeding on, for changes in diet should be gradual.

When a ewe is about to lamb, she will often detach herself from the flock and go into a hedgerow, or the corner of a field. Indoors, she will likewise tend to go into a corner. She will scratch a bed and adopt a straining posture, holding her head up high in the air. Sometimes ewes, like women, experience 'false alarms,' but generally this behaviour is followed by the appearance of a water bag - the protrusion of the amniotic membrane - from the vulva. Leave her alone at this stage, and come back in half an hour or so. In most cases she will have delivered without problems, and will be busy licking her lamb or lambs. The afterbirth should follow on its own, later. - If nothing has happened, wait a little longer. The water bag may burst and hang down like a piece of gelatinous string. Wash your hands and forearms in soapy water, and smear them with antiseptic jelly. Look, or feel inside. If the presentation is normal you should see, or feel, a nose and two front feet (fig. 4-5,A). Grip a foreleg and pull gently, then do the same with the other. Pull the legs alternately so that the shoulder blades will be eased through the narrowest point, one at a time; when the head appears, pull downwards. With most normal presentations you should have little difficulty in delivering the lamb. A very large lamb, or one with a relatively large head, can be difficult and requires patience. Always check that a newly-born lamb's nostrils are clear, and that it is breathing; if not, try blowing into its mouth, slapping its ribs, and moving its forelegs to and fro.

A proportion of sheep show abnormal presentations, which require assisted birth. They may be listed as follows:

Head back. (Fig. 4-5, B). Sometimes you will feel two forelegs, but no nose. You should slide your hand in, find the head, and carefully bring it forwards, proceeding then as with a normal presentation. Sometimes the head will not stay in place, and a loop of cord has to be passed behind the ears and in front of the mouth. Pull the cord as the legs are eased forwards. Never exert any pressure on a lamb's lower jaw, for it is fragile and easily broken. Cord loops should be sterilized each time, before use.

One leg back. (Fig. 4-5, C). Sometimes you will feel the nose, but only one foreleg. You should ease the head backwards and slide your hand in, finding the other foreleg. Hook a finger round the middle of the leg, and bring it forwards, cupping the hoof in your hand to prevent it from damaging the uterine wall. You can then proceed as with a normal presentation.

Both legs back. Sometimes only the head is presented forwards. In this case, proceed as above, dealing with one leg at a time.

Hind legs first. (Fig. 4-5, D). Sometimes the hind legs are presented. Pull the lamb out quickly, for, as soon as the umbilical cord breaks the lamb will start to breathe, with a risk of drowning in the uterine fluid.

Rump first (breech). (Fig. 4-5, E). Sometimes only the rump and tail can be felt, the hind legs sticking forwards along the inside of the uterine wall. Some say that the whole lamb should be turned round, which is difficult and may result in severing of the umbilical cord; others say that the best way is to bring the hind legs back, taking care not to damage the uterine wall, and then deliver quickly backwards. The inexperienced may need help with this presentation, either from a neighbour or the vet.

Entanglements. In multiple births the lambs sometimes become entangled. It is not easy to sort out which leg belongs to which lamb. They must be separated, and one of them delivered whilst the others are pushed back. Often the second, and third in the case of triplets, will follow without assistance. In sorting out legs by feel, it helps if you remember that the lamb's fore and hind legs are hinged differently, so it is easy to know which you are dealing with. If in doubt, you should call the vet.

Occasionally a ewe's cervix fails to dilate. You should call the vet., or, after telephoning, take the ewe to him. Sometimes, before, or after lambing, a prolapse of the vagina may occur. This should be pushed back and held in place by inserting a special plastic prolapse retainer, which can be tied to the sheep's wool

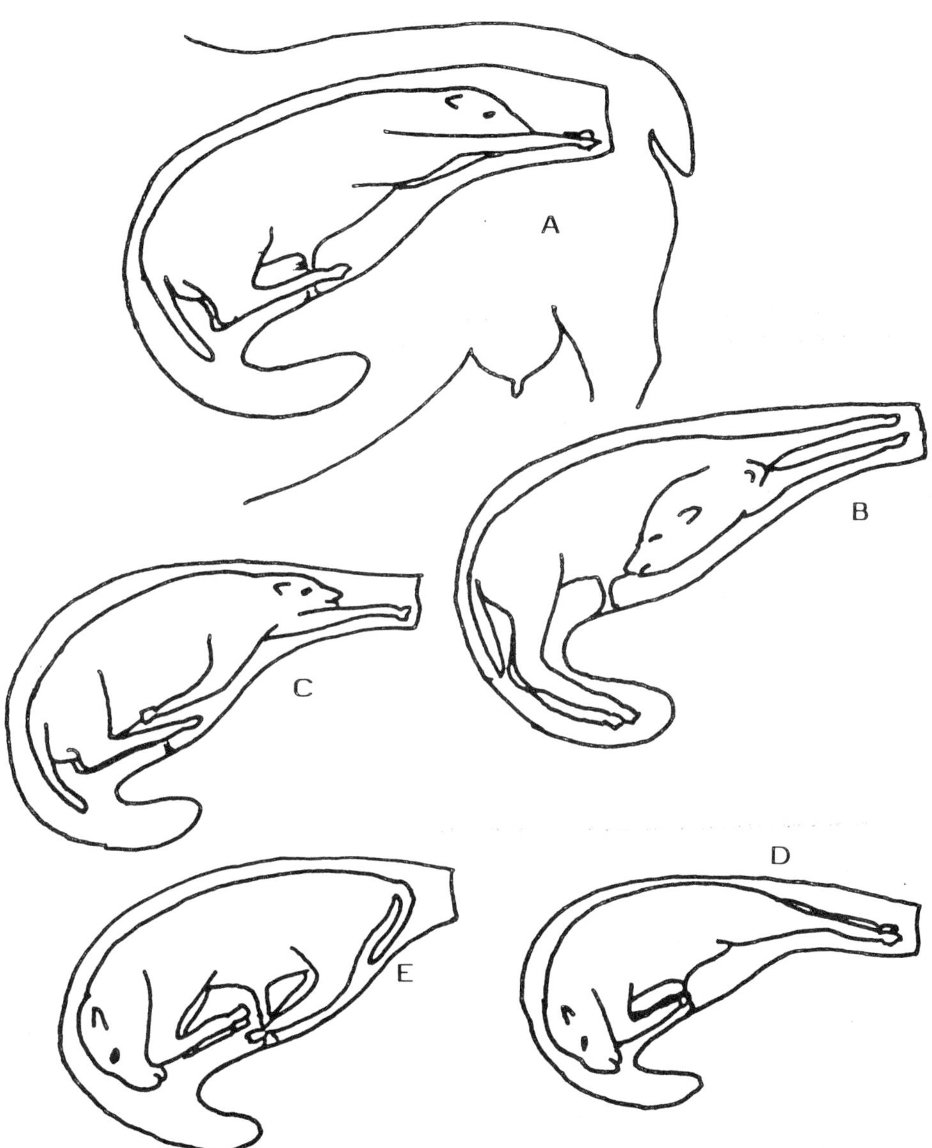

Fig. 4-5. Lamb presentations. (A) Normal; (B) head back; (C) one leg back; (D) hind legs first; (E) rump first (breech).

either side of the tail. In emergency, a piece of stiff wire bent in a long U-shape, 14cm x 4cm, can be inserted, and twine attached to small loops in the protruding ends can be tied to the wool. A ewe with a prolapsed vagina is likely to prolapse again, next season; it is best to sell her for meat when her lambs have been weaned. After giving birth, a ewe may have a prolapsed uterus; the vet. should be called, for returning the prolapse is a delicate operation; the ewe must be held bottom-up, and the uterus carefully and slowly returned by delicate manipulation of soapy hands. Afterwards a few stitches must be inserted to prevent a recurrence, and anti-biotics must be injected. The ewe should be sold for meat when her lambs have been weaned.

Unless you are unlucky, abnormal presentations will occur only in a small percentage of your flock. However this percentage is signifi-cant in relation to your profit margin. Losses from abnormal presentations often occur because they have not been noticed early enough. If left too long, half-delivered lambs will begin to swell from the pressure exerted on them; their eventual delivery may be fatal; half-extruded, entangled twins may become so swollen that one has to be cut up to permit delivery. Furthermore, if a ewe loses her lamb(s) she is likely to be barren in future, unless you can persuade her to foster an orphan, or spare twin lamb. To help per-suade her you can smear the candidate lamb with her dead lamb, or skin the latter and fit it like a jacket on the former, holding the mother in an adoption crate (fig. 4-6), if necesary. To avoid the worst of these problems, frequent and careful surveillance is the answer, and that is why we recommend lambing inside, if at all possible.

Once the lamb, or lambs have been born, or delivered, the mother will turn round and lick them. The afterbirth will follow; it should be disposed of if the mother does not eat it. The lambs should soon start to suckle, unless they are weak or sickly. You should spray each lamb's navel with iodine or other antiseptic, to prevent infec-tion. For the next day or so it is a good idea to keep mother and lamb(s) in a small pen made of four hurdles (fig. 4-6) with water, hay, and concentrates. This gives the mother a chance to get a good feed, and it protects the lambs from predators when they are at the most vulnerable stage. It also enables the mother to come to know her lamb(s), avoiding problems of mis-mothering. Check that the lambs are feeding - if in doubt check that both the mother's teats are giving milk. If a ewe is only 'sparking on one cylinder,' she will probably only be able to rear a single lamb, and any second, or third lambs should be fostered, or hand reared. If you have had your hand in the mother's uterus to assist delivery, she should be given an antibiotic injection to prevent uterine infections

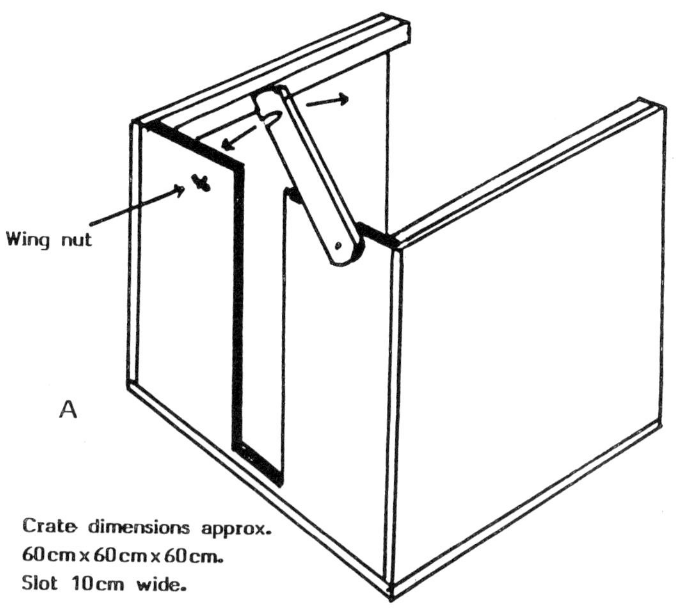

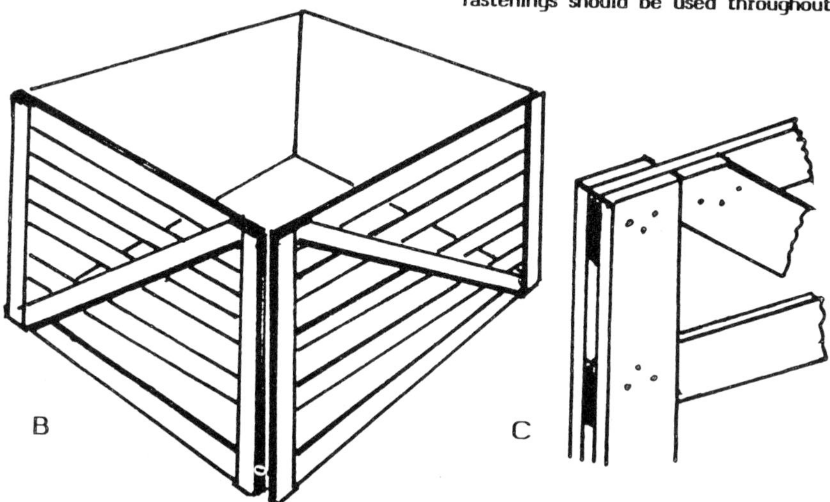

Fig. 4-6. (A) Adoption crate. The crate should be weighted or fixed in place. The ewe's head goes inside the crate, her body outside. (B) Hurdles used to make a temporary pen for ewe and lamb(s). (C) Detail of hurdle construction.

which could make her barren. At this stage the lambs may be tailed by use of a rubber ring. The ring should not be placed too high up - it is illegal to shorten the tail above a ewe's vulva or a ram's anus. This operation should be done within seven days of birth; you will need an elastrator for stretching and fitting the rubber rings. Ram lambs may also be castrated, up to two days old, using rubber rings, but this causes them to suffer until numbness sets in. We prefer to use a bloodless 'Birdizzo' castrator later, before they are six weeks old. The Birdizzo pinches the spermatic cord without cutting the scrotal sac; only each edge of the sac, containing a spermatic cord, should be pinched, so that a blood supply is left in the rest of the scrotum.

If, despite all your efforts, you end up with some orphan lambs, you can keep them collectively in a small pen. Feed them at first by hand, using a bottle with a lamb teat. If any have not had a first feed of colostrum from their mother, they should be given it for their first two feeds. Cow's colostrum is all right, and it can be stored in deep freeze. An inferior artificial substitute can be made by mixing 2.5ml glucose, 2.5ml cod-liver oil, a raw egg, and 150ml milk. Lambs should not be reared on cow's milk, for, although they thrive for a time, they eventually die. They should be reared on orphan lamb food, which is sold in powder form, to be mixed with water before use. When your orphan lambs are a few days old, they may be provided with a lamb bar - you can purchase special lamb teats which can be arranged so that several lambs can feed from liquid in a single container.

You should spray numbers on the sides of your ewes and lambs, using the same number for mother and her lamb(s), with a different colour for singles and twins. This is useful for keeping records, and it prevents mistakes if you sell any ewes with their lambs. Try, if possible, to raise your lambs on clean, or safe pasture. If they are on contaminated pasture they will need worming more frequently. A pot-bellied shape is often a sign of Nematodirus infection. Like your ewes, your lambs can be protected against several diseases, in one vaccination.

If you wean your lambs whilst their mothers are still giving milk, the latter should be kept inside, or in a yard, under starvation conditions for twenty-four hours, and then put on poor grazing. After three days, their udders should be checked. If any are distended they should be milked to ease, and checked again three days later; this helps prevent mastitis (udder infection). The ewes may be kept on indifferent pasture for a time, but they should be got in good condition before tupping.

CHAPTER 5. CATTLE.

Even before the advent of milk quotas, the trend for many years had been for dairy herds to become larger, the smaller dairy farmers gradually disappearing, so that it would have been unwise for anyone to think of starting dairying in a small way. Now that the quota system is in operation, it is impossible to make a start unless you can find a small dairy farm for sale, with a quota.

Cattle options for the majority of smallholders are restricted to keeping one or two house cows, a small suckler herd, calf rearing, and raising beef cattle (generally to be sold as stores at twelve to twenty months). One other cattle option, raising dairy heifers, is risky because most dairy farmers rear their own replacements. Therefore many of the dairy heifer calves sold at mart are not of the best quality, and they include some freemartins which are unfit for breeding.

Handling.

You will need a cattle crush to hold a cow, or any other cattle you keep, for purposes of veterinary inspection and routine treatment. The latter includes worming in spring, a few weeks after turn-out, and again when brought in for the winter; treatment against warble fly around October by applying a measured systemic dressing to the animal's back, or by an appropriate injection; injecting against blackleg if prevalent in your area, and testing for tuberculosis and brucellosis by a vet who will be sent by the Ministry of Agriculture.

The simplest form of cattle crush can be made from two farm gates fixed to pen in the animal. The gates must be strong and securely fixed, for, should one fall over, you, or your cow could break a leg. It is safer to have a purpose-made cattle crush. Metal ones may be purchased, but they are fairly expensive; if you have only a few cattle, a cattle crush can be made using stout posts of about 150mm x 100mm with rails of 100mm x 50mm, and stout doors either end (fig. 5-1A). The posts should be set in concrete in the ground; in fact the entire crush should be on a concrete base, if possible. The timber rails should be securely fastened to the inside of the posts, and protruding bolt or nail heads should be avoided. Cross pieces linking the tops of the corner posts help to give rigidity to the structure. One or two farm gates can be fixed temporarily to form a funnel-shaped entrance.

Large cattle can be haltered when you wish to lead them, as shown in fig. 5-1B. For calves, a simpler form of rope halter may be used, as a temporary measure (fig. 5-1C).

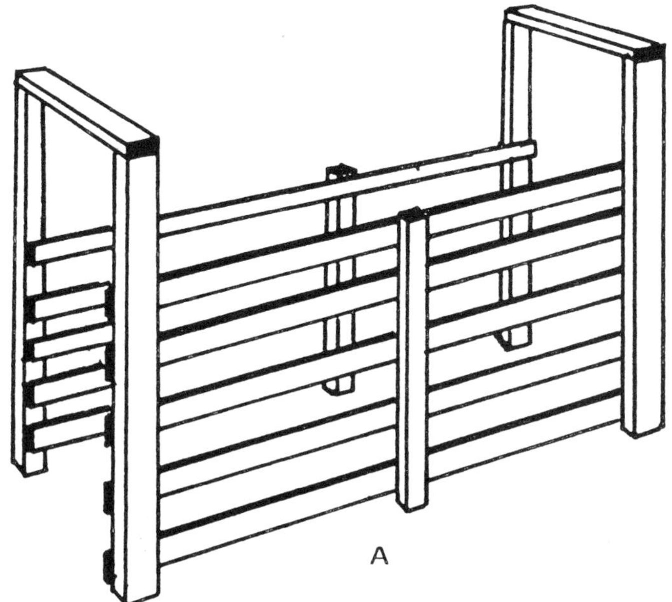

Basic construction of cattle crush. Stout doors should be added each end.

Rope halters should be placed over the heads and behind the ears first, then over the nose and under the lower jaw.

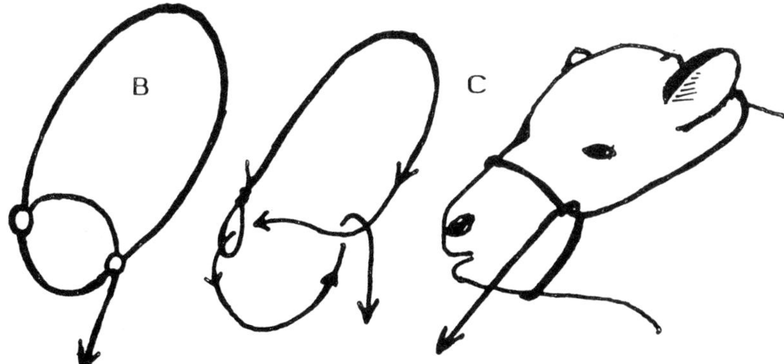

Fig. 5-1. (A) Timber cattle crush. (B) Rope halter. (C) Rope with knotted loop used as temporary halter for a calf.

Feeding, etc.

In this section we deal with cows, growing calves, and store cattle. Young calves are dealt with later, under a separate heading.

Grass, in the growing season of spring and summer, and good hay in winter, will maintain a cow. During the early flush of spring grass she will only need a handful of concentrate feed as a 'reward' for coming in. For most of the year, however, you will have to feed her manufactured concentrates if you want a reasonable output of milk. The best time to feed concentrates is when she is tethered for milking; the precise quantity depends on the particular cow, and the amount of milk required. If you feed grain, instead of manufactured concentrates, be sure to provide a mineral lick; you should seek expert advice if you wish to mix your own feed to be sure of a correct nutritional balance. During the growing season your cow will be kept outside on grass, and brought in twice a day for milking. A cow milks 'off her back,' that is to say, from her stored reserves, for the first six weeks after calving. Therefore, to obtain a good milk yield for yourself and her calf, she should be kept in good condition. Your cow should be fed well for the three months preceding the birth as this is the time when the growing foetus makes demands on her. See that she has plenty of good grass in summer, or concentrates and hay in winter. If she is at all thin, add barley to the ration, and extra concentrates. If you are in any doubt, it is our experience that other farmers are helpful and willing to advise when asked.

Growing calves, whose rumens have developed, and older store cattle should thrive on grass alone in spring and summer. Once again, mineral licks should be provided. Hay or silage, plus a daily concentrate ration should keep them growing in winter.

Whatever your cattle enterprise, if it involves you in keeping cattle through the winter, the number you can keep will probably depend on the size of your hay or silage crop. There will also be the expense of concentrate feeds to consider. During winter, your cattle may be permanently housed, either loose on deep litter, in cubicles arranged so that the cattles' bottoms stick out and do not foul the litter, or tethered. Cattle tethered, or in cubicles, need mucking out daily.

In-wintered cattle should have a supply of clean drinking water, hay, feed-straw, or silage at all times, and a daily ration of concentrates. Drinking water can be made available from automatic cattle drinkers; sometimes the cattle have to be taught how to operate them, but generally they learn quickly. All food should be

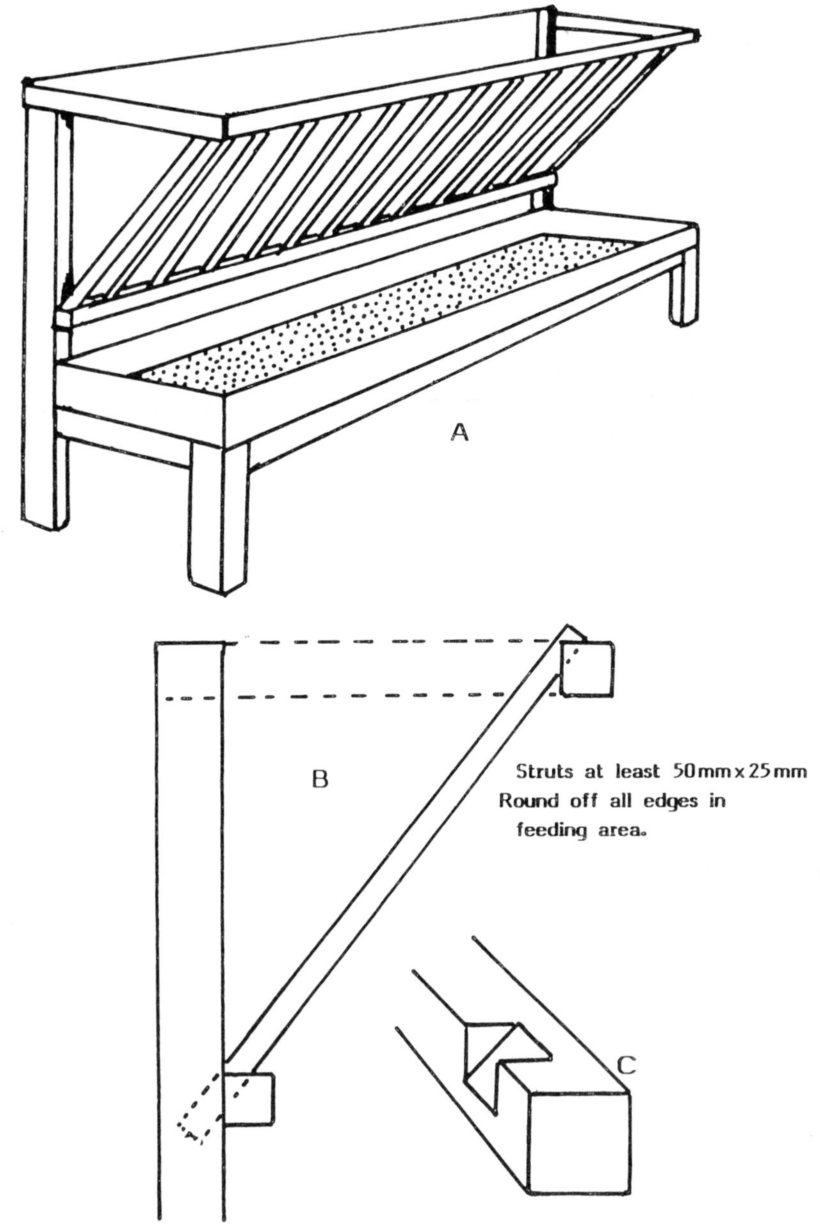

Fig. 5-2. (A) Combined hay rack and feed box. (B) and (C) details of construction. The top of the feed box should be 60 cm above ground.

brought to tethered cattle and placed so that they can reach it. Untethered cattle are generally provided with inclined hay racks with a feed box below. The feed box prevents wastage of the hay, or feed straw, and also serves to hold concentrates, which the cattle clear up quickly. Combined hay racks and feed boxes may be purchased, made of galvanized steel or wood, or they may be constructed on the farm (fig. 5-2). Cattle in-wintered on deep litter, or in cubicles, may also be fed on silage. This may be brought to them daily, or they may be given direct access to the silage pit. In the latter case, a set of 'tombstones' - wide wooden uprights - obliges them to feed separately, without annoying one another. The tombstones are moved forwards as the silage is eaten up.

If cattle are kept out in all but the worst winter weather, they risk poaching the land, especially on clay soils, and their feed will cost more because of the extra energy used in generating body heat. Furthermore, winter feed has to be taken to them, which can mean hard work in inclement weather. A compromise that one of us has found to be successful, is to rail off part of a barn where they can be provided with deep litter, water and feed, but from which they have free access during the day to the great outdoors. This system uses less bedding straw and ensures that the cattle are hardy and suffer no checks when turned out to graze in the spring.

Calving.

At least six weeks before her next calf is due, the cow must be 'dried off' (milk production in the udder must stop). Do not milk her for a week, then strip (milk) her right out, and put a tube of dry-cow therapy up each teat. This may be bought at a farming supply shop, or from a vet. It is essential as a precaution against mastitis.

When your cow is due to calve, her bag will be big and very tight; then the pin bones each side of her tail will drop; she will become restless and go off on her own. In a normal calving you should neither help nor interfere; your cow will start to have contractions and a bladder will come out of the vulva, increasing in size until it bursts. By this stage your cow may lie down, and the calf's forefeet may appear, surrounded by a membrane which will soon burst. If the two feet do not appear first, with the nose a few inches behind, call your vet. In a normal birth, the calf should be delivered after a pause - all you will need to do is to see that the calf's nostrils are clear. The mother will clean the calf by licking it. The afterbirth should be left to come out naturally; its own weight helps to draw out the remaining parts gradually.

The milk produced during the first four days after calving is colostrum; this is valuable to the calf, for it confers immunity against several diseases. Colostrum is not suitable for domestic use; it may be frozen in small plastic containers, and kept thus as a first feed for any orphan lambs, or, should you be unlucky, an orphan calf. Whatever happens, your cow should be milked during the first few days, for your calf may scour if it takes too much milk at the beginning; also her milk yield will drop if she is not milked at this time. During the first three months the calf will need most of the milk, but dairy breeds always have some surplus during this period; after that time you may take a greater proportion of the yield.

Mating; artificial insemination (A.I.)

Six weeks after calving the cow will come 'bulling' (into oestrus) and thereafter at twenty-one day intervals. The gestation period is nine months, and this enables you to plan future calving dates.

If you don't wish to keep a bull, which is too costly for less than thirty cows, you can use the A I service. A cow stays in oestrus (on heat) only for a few hours, so you must learn to tell when a cow is bulling, and 'phone the A I service immediately. When cows are in oestrus, they mount one another. If a cow is being mounted and she stands still, she is bulling; if she moves away, the cow that mounted her is bulling. If you have only one cow, you should observe her closely around the period of her 'due' days for slight changes in behaviour, such as restlessness. She may show a drop in milk yield, and she may have a slight discharge from the vulva, but not always. After 'phoning the A I service, you should bring your cow in, tether her, and prepare a bucket of hot water, soap, and a towel ready for the A I man when he arrives - remember that he may have many other calls to make the same day, and that cows only stay in oestrus a short time. As an alternative to watching your cattle for signs of bulling, you can buy 'heat-detection strips' to stick on the back of your cow; they are expensive and unnecessary since any good farmer watches the cattle for a few minutes night and morning, anyway, to check for signs of illness, etc.

The house cow.

Many of the books on self sufficiency deal with the house cow in detail. In this book, however, we assume that you will wish to keep your house cow on as commercial a basis as possible, so that her calf will pay for her keep and provide a small profit when sold as a store. This means avoiding rare breeds whose progeny will be difficult to sell, and Channel Island breeds whose calves are worthless as beef. To make a profit, your cow should be inseminated with a 'beef' breed such as Hereford, Angus, Charolais, etc. This should qualify your house cow for the suckler cow premium. The house cow should be a commercial breed such as Friesian, or perhaps an Ayrshire, Welsh Black, or Dairy Shorthorn if they are still popular breeds in your area.

After calving a cow will continue to give milk for up to nine months. If you wish to have a continual supply of milk, you will need two cows phased to calve at different times of the year; otherwise you will have to buy milk for about two months each year.

We would recommend your consulting a friendly neighbouring farmer to assist you in buying your cow. A house cow may be purchased in a farm sale as a calf, or older, as a bulling heifer, reared on, mated by A I, calved, and got into milk. This is a slow process and we would not recommend it to a beginner, as you will both be learning to milk at the same time. It is also no cheaper than buying a freshly calved cow, when all the rearing costs are taken into consideration. If you are inexperienced, you should think of buying a proven, docile house cow that has already calved at least once, and is guaranteed to milk by hand. The price you will have to pay will, of course, depend on the age and breed of the cow. Sometimes it is possible to purchase, relatively cheaply, an old, but sound cow rejected from a dairy herd because she has a low yield. She may be nearing the end of her useful commercial life and be sold just for her meat value. Such an animal may give you a year or two of service, but if you don't want to lose money, you should eventually sell her for beef before she dies on you or begins to cost you a fortune in vet's fees. However, if you can afford it, we would recommend you to purchase a house cow with a calf. You will get some of your money back later, when the calf is sold. A lactating cow must be milked twice a day, whether you want a holiday or are lying at death's door. Her growing calf, given half a chance, will take all the milk, so that in an emergency milking will not be essential. If you propose to milk by hand, look for a cow with big teats; tiny teats are difficult for a beginner, and hard work even for an expert.

Milking.

A seamless stainless steel bucket is best for collecting the milk, for it can be kept thoroughly clean. There are machines - 'bucket units' - for milking one cow at a time, with rubber-lined cups which fit on the teats. If you have only one or two cows, the business end of this equipment will have to be cleaned out each time you milk, so there is not such an advantage, timewise, over hand milking, unless you have a high yielder without a calf. The choice is one of personal preference, but hand milking half a gallon from a cow with a calf takes less time than assembling and cleaning a bucket unit, and saves an expensive capital outlay. Even if you are given a unit, all its rubber parts have to be replaced every six months. You should milk twice a day, to a regular time-table, without variation. If you milk by hand, you will need a three-legged stool, and you should milk from the cow's right side. Always speak when you approach her, place your hand on her back and run it down her side before sitting down with your head against the cow (so that she knows where you are). Wash the teats and udder, and dry them with a disposable paper towel. Put the bucket in place and grasp a teat in each hand. Using your hands alternately, contract your fingers in succession, the upper one first. It is best to get someone experienced to teach, and watch you, for the first few times. After milking, dip each teat in a proprietary teat dip, as a precaution against mastitis (udder infection).

The suckler herd.

An experienced farmer once said to one of us that 'a single-suckler herd is one of the most pleasant, but least profitable, ways of livestock farming.' We have, however, pointed out in an earlier chapter that there are advantages in keeping cattle along with sheep, to give a safe grazing strategy and to help keep the land sweet. A small suckler herd can therefore be considered as a way of supporting a successful sheep enterprise. The main expenses are the cost of feed, and provision of winter housing for those breeds not hardy enough to be out-wintered, or on wet land that will be poached (turned into a quagmire) in winter.

Any breed of cow will rear a calf, but from a commercial point of view, some breeds are quite unsuitable as suckler cows. For example, Channel Island breeds, although excellent as dairy cattle, are unacceptable for meat because of their yellow fat. Hereford-Friesian cross cows are ideal, also Fresian, although the latter take more feeding and must be housed in winter. Breeds such as the Angus and Welsh Black are hardy and can be out-wintered, but they are relatively slow growing and, after their first calf, they should be mated with a commercial fast growing beef bull such as a Charolais. Other suckler cows are best mated with a beef breed, such as Angus, Hereford, Charolais, etc. The Angus, being small, is often used for first calvers. If you mate suckler cows with a dairy bull, this may render the enterprise ineligible for subsidies which, in some areas, are high enough to be a definite factor in the profit margin.

As an alternative to single suckling, multiple suckling may be considered. In this case a large milky cow is needed. Sometimes docile, elderly Friesian cows taken from a milking herd are used. One of the problems is that the nurse cow must be one prepared to accept the bought-in calves. This problem can be largely overcome by keeping a pair of calves in each of two pens, separated by a narrow passage in which the nurse cow can be yoked whilst she feeds. A door in each pen can be lifted, giving the four calves access to suckle the yoked cow without risk of being kicked. This arrangement is shown in fig. 5-3 A.

Calf rearing.

Your calves can be bought at auction marts, or privately from local farmers. It is illegal to sell them immediately they have been born; this is to ensure that they will be sold after they have had their first feeds of colostrum from their mother, which gives them immunity against several diseases. Apart from making sure that your calves have had colostrum, you should also check that they have been ear tagged, or tattooed in the case of pedigrees, with the herd number of the farm on which they were born; this is also a legal requirement. You also are required to keep an animal movement record book in which you enter the herd numbers of all cattle purchased, and whom, and where they have been purchased from. You should also record the sale, and movement, of any cattle off your holding. This requirement holds good for other farm animals, such as sheep, except that they do not have to be tagged with a herd number.

Generally speaking calves cost more when bought privately, but some busy dairy farmers are prepared to sell indifferent calves fairly cheaply, to save themselves the trouble of taking them to market. Obviously, calves bought off a farm have run less risk of contracting disease, but at an auction mart you have a bigger choice, and sometimes a better chance of picking up a bargain. Remember that it is easy to become 'carried away' at a mart, and that cheap calves will cost you just as much to rear as ones of good quality. Some calves, such as bulls of the Jersey, Gurnsey, and Ayrshire breeds, are barely worth the cost of rearing, for they have little meat value.

Calf housing should be light, airy, well-ventilated and free from draughts. A building with solid walls up to about 1.8m, followed by space boarding all round, is ideal. Gently sloping concrete floors, with drainage gulleys situated outside the calf pens, are an asset; otherwise enough gravel should be put down to prevent the litter from becoming waterlogged. Floors and pens should be thoroughly disinfected using a recommended product, or steam cleaned,

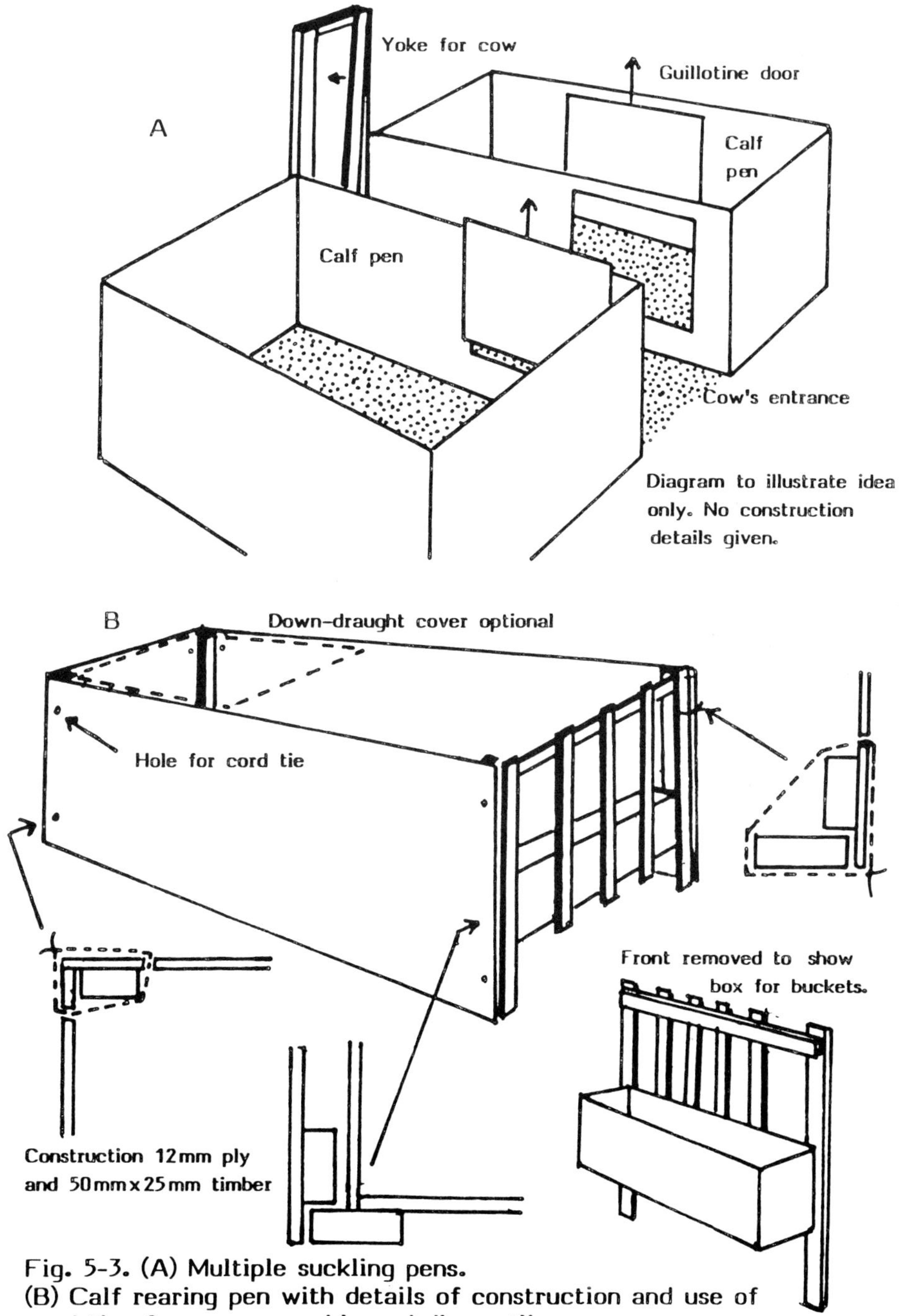

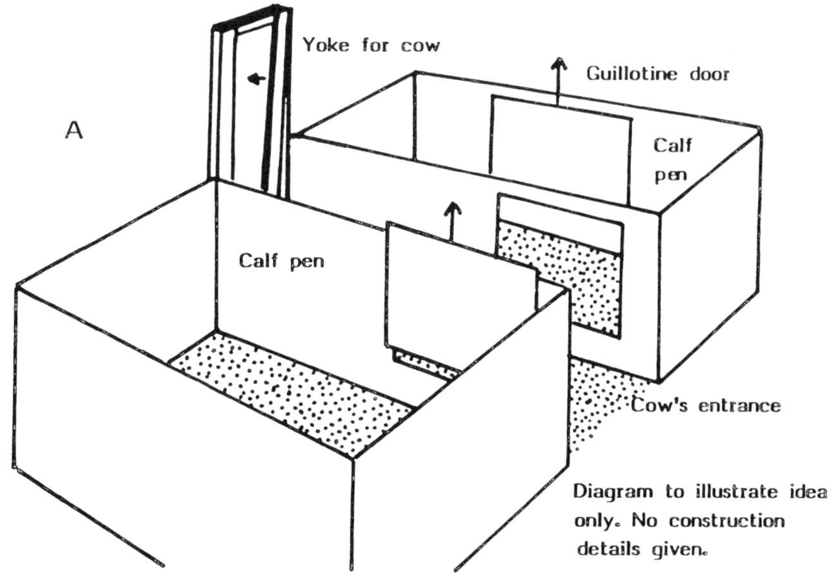

Fig. 5-3. (A) Multiple suckling pens.
(B) Calf rearing pen with details of construction and use of cord ties for easy assembly and dismantling.

CHAPTER 6. POULTRY.

Not all poultry ventures are successful. If you intend to set up an intensive battery, or deep litter, poultry enterprise, you will have to make a large initial capital outlay. As a smallholder you will probably not wish to become involved in intensive kinds of factory farming. You can start, therefore, in a small way, exploring the market potential, and checking the economics of the venture at the same time. You can expand later, after having verified the cost of your poultry keeping, and having established some outlets. It is important not to become carried away in the early stages and collect a great assortment of poultry, willy nilly, hoping that they will give you some sort of return. Such an approach leads to ever increasing weekly feed bills, and may lose you more money than you think. If you are thinking of selling dressed table poultry, remember that you may have to comply with regulations relating to killing, eviscerating, packing, storing, and transport of your products.

In the following sections we comment separately on different kinds of poultry, followed by general accounts of breeding, incubation, rearing, housing, feeding, and general management.

Chickens.

In large, intensive enterprises, chickens are kept either for egg production, or for meat. On a smallholding scale you may select breeds suitable for egg production alone. We would recommend them to those who do not relish the thought of killing their own cockerels for the table. Alternatively you may keep the heavier breeds of dual-purpose birds; they give a reasonable output of large eggs, and the cockerels are suitable for fattening. There are also modern table breeds which you can purchase as day-olds and rear for home consumption, or sale. Some of these modern table breeds, such as the Ross, have a phenomenal growth rate and can be quickly and cheaply got to table weight. Generally speaking, day-old chicks travel by rail without problems.

You may opt for keeping just a small laying flock to supply your domestic needs, with the occasional surplus going to neighbours. Older breeds, such as the White Leghorn, are suitable for this purpose. An alternative is to purchase, quite cheaply, year-old ex-battery hens. A few of them may die on you, but the rest may find a new lease of life in the improved environment you will offer them. If you want also to fatten a few birds for the table, hatching some eggs under a broody hen or in a small incubator, you could keep a small flock of a dual-purpose breed such as Rhode Island Red, Light Sussex, Maran, etc.

If, however, you wish your poultry venture to have possibilities for commercial expansion, you should think in terms of specializing in selling eggs, breeding stock, or dressed table birds.

Eggs. Nowadays there is a significant demand for free-range eggs, and some people in the cities are prepared to pay a good price for them. The demand is for a large, brown-shelled egg, as laid by many of the older dual-purpose breeds, and some of the modern brown egg strains. Starting small, building up gradually, you may be able to market your eggs locally. If you wish to expand further, you should look for a distributor to market your eggs in towns and cities where the price is higher. If, before buying a holding, you are really set on this kind of enterprise, you should first ensure that suitable distribution and marketing facilities are available in the area.

Breeding stock. Modern strains are hatched in enormous quantities in hatcheries which supply the intensive poultry industry. As a smallholder you can think only of breeding livestock to be sold to farmers, smallholders, and backyarders. A reasonable start can be made by crossing a Rhode Island Red cock with Light Sussex hens. This is an autosexing cross, that is to say that the sex of the chicks is easy to distinguish. The female chicks hatched are brownish, the males yellowish. This is useful as many of your customers may wish to buy pullet chicks only, for egg production. The male chicks may be sold a little cheaper, or reared to table weight. These RIR x LS hybrids were once popular and commercial; they are still remembered and should still be saleable. Alternatively, you could make a start with one or more pure breeds. Your chicks will not be autosexing, but you may find someone to teach you how to sex day-olds by opening the vent. One problem with pure breeds is that many of your customers will be wanting to breed from your livestock, and they will ask for pullets with an unrelated cockerel. This means that for each pure breed you keep, you may need two separate pens, each with its own cock.

Table birds. You can use any of the heavy dual-purpose breeds for producing table birds, and this activity may be combined with egg production, or sale of pullets. Alternatively, you can buy modern fast-growing hybrids as day-olds, and rear them to killing weight. Although they have to be bought in, the labour and cost of keeping your own breeding flock is eliminated. Given sufficient food, strains like the Ross can gain an average of a pound in weight per week, and they can quickly be got to quite heavy finished weights. They are best reared inside, fairly quickly, for older birds sometimes become unsteady on their feet; this means that they may not be to everyone's liking. They are ideal for selling at festive and holiday seasons, such as Christmas and Easter. The fact that you

purchase these birds as day-olds gives you flexibility in the number
you rear at a given time of the year; your own fixed-size breeding
flock could not offer the same flexibility in terms of quantities
of hatching eggs.

Ducks.

Ducks are wonderful birds to keep. Timid, docile, and easily upset
by strangers, they are best kept in a peaceful setting. Although
they can be kept away from water, they thoroughly enjoy having
access to a pond or stream. The large breeds mate more successfully on water; on land they risk rupturing themselves. If you have
neither pond nor stream, a pond can be dug, lined with plastic
sheeting, and rendered in cement (because the ducks' feet can
puncture a plastic lining). There should be a non-slip ramp at one
end where the ducks can climb out. An easy way to keep your pond
supplied with water is to run a flexible pipe from a container
fitted under a downspout on your house, or outbuilding. Commercial
breeds of ducks rarely go broody, and they do not make good
mothers. Eggs may be hatched in an incubator, and the ducklings
brooded artificially. A broody hen will sit a few duck eggs and
become a good foster mother to the ducklings; she will try to teach
them how to scratch for food, and will appear quite perplexed if
they go for a swim.

There are egg-laying, dual-purpose, and table ducks. Only table
ducks have commercial potential.

Egg-laying breeds. Years ago, the Indian Runner was the preferred
egg layer. This has largely been replaced by the Khaki Campbell
and Welsh Harlequin breeds. When buying stock, be careful to buy
from a breeder whose ducks have a high egg output; many flocks
have been left unselected and ducks from them can have poor egg
production. If you obtain a high egg strain of duck, your eggs may
be produced at less cost than chicken eggs. Because of fear of
Salmonella poisoning, duck eggs do not sell easily in the UK. This
means that you should keep egg-laying breeds for your own use, and
that of a few friends and neighbours who may have a preference
for them, or use them for baking. Duck eggs, in themselves, are
no more prone to Salmonella infection than are chicken eggs; the
greater risk arises because ducks are not fussy where they drop
their eggs, often laying them in wet, dirty places. If you wash duck
eggs, be sure to use water at a slightly higher temperature than
the eggs; this causes the air in the shell to expand, pushing bacteria
outwards, rather than drawing them in.

Dual-purpose breeds. Some people consider the Welsh Harlequin as a dual-purpose bird. In our experience, much of the Welsh Harlequin breeding stock now available does not produce birds heavy enough for most table purposes. The Whalesbury breed, which was obtained by crossing a Welsh Harlequin drake with Aylesbury ducks, was a better dual-purpose bird; at one time it had some commercial potential. We have never seen Whalesbury breeding stock for sale. We attempted to repeat the cross, without success, but we obtained interesting results from the reciprocal cross, Aylesbury x Welsh Harlequin. Some of the progeny were black and white, others had a most attractive mallard-like appearance; they grew reasonably fast to an acceptable table weight, and were good layers. For those who wish to have a go at cross-breeding, we would add that one of our neighbours reared Aylesbury x Khaki Campbell hybrids only to find that they were highly strung and prone to heart attacks.

Table breeds. Years ago, the Aylesbury was the commercial breed of table bird in the UK; in the USA it was the Pekin. Many of the modern commercial strains of ducks contain Aylesbury and Pekin blood, and they have a slightly improved egg output. These modern strains have been produced to provide a fast growing duckling that can be killed as a medium priced table bird at eight or nine weeks old. After about nine weeks they go into the moult, and are difficult to pluck. A few producers kill their birds after the moult, at about twelve to fourteen weeks. This gives a heavier bird, but it is relatively more expensive, for ducks become less efficient food converters as they grow older. If you want to breed table ducks for profit, be sure to obtain a modern commercial strain. If you buy 'Aylesburys' from a local farmer or smallholder, you may be lucky, or you may find you have bought only white ducks of doubtful quality.

If you want to keep table ducks for your own consumption, without much regard for economics, you may consider Rouens or Black Cayugas. The Rouen is an attractive, heavyweight bird, but it is not prolific. Its meat is renowned for its flavour. Black Cayugas are, in fact, an iridescent beetle green in colour; both these breeds, along with many others, are suitable for ornamental purposes.

Geese.

Before the turkey became popular, the goose was the tradition Christmas fare. There is a small, but significant demand for dres goose in the United Kingdom, and good prices can be obtained this up-market product. A limited potential exists, therefore those prepared to go to the trouble of setting up a comm breeding enterprise.

Because goose production has been in decline in the UK for many decades, many of our 'barnyard' geese are of doubtful ancestry, probably inbred, and quite unselected for egg production and conformation. You should be careful, therefore, where, and from whom, you buy your geese.

The requirements for commercial goose production are maximum prolificity coupled with a fairly heavy finished carcass. The larger, heavier breeds available in Britain, such as the Toulouse and Embden, are not very prolific. Smaller breeds such as the Roman and Chinese, lay more eggs, but have a lower carcass weight. The would-be commercial breeder is recommended, therefore, to start by crossing the larger, with the smaller breeds, keeping the female hybrids for his or her breeding flock. Ganders of one of the large breeds should then be mated with the hybrid females. The result should be a reasonably prolific flock producing fairly large table offspring. This could be a slow process, for fertility in geese can be poor in the first season. Furthermore, careful planning and management would be required because geese mate for life, and sets would have to be made up in temporary pens in which a gander could settle down with four or five geese. Later on, several sets could be run together, if required, but the initial segregation would be necessary to prevent dominant ganders from collecting more geese than they could fertilize. Very little selective breeding has been carried out on geese; those who would be prepared to take the trouble of ringing or tagging, and keeping proper records, could well be able to improve egg output and body conformation. - If you do not wish to involve yourself in so complicated a breeding programme, an improved goose flock can be produced simply by crossing two breeds together - cross-bred geese often benefit from hybrid vigour.

If you just want to keep a few geese, we recommend the buff kind. Relatively speaking they are quieter and more docile than other geese, and they are most attractive. Of medium weight, they can make an acceptable table bird. Pure Brecon Buffs have pink bills and feet. Many geese sold as 'Brecon Buffs' have orange bills and feet; strictly speaking they should simply be called buff geese. We doubt if the colour of the bill and feet has much relationship to performance characteristics.

Geese are grazing animals. On good, short grass, they should only need feeding in autumn and winter. Good grazing is the key to profit in any goose enterprise. If you are thinking seriously of 'going into geese,' we recommend you to consider first re-seeding their intended pasture, mowing the grass when it is established, *for only good quality, short grass will give satisfactory results.*

Muscovies.

Muscovies are a favoured table bird in Australia. Their meat is of excellent quality, with a slightly 'gamey' flavour. The commercial duck industry in the UK has never exploited them, perhaps because of their restricted egg output, and their flightiness. Although often described as ducks, they behave more like geese, for they are grazing birds. Like geese, they will sit their own eggs. They offer real possibilities to smallholders who may find up-market outlets in restaurants, hotels, and high class butchers. Their flightiness can make management a problem. Their wings can be clipped, but you should note that it is illegal to clip the wings of free-range birds, and they would have to be reared in pens, protected from predators. Like geese, to be economical, they must be provided with good, short grass on which to feed; otherwise your feed bills will take all the profit.

Guinea-fowl.

A few Guinea-fowl may be kept to range freely outside the house, as 'watch dogs,' and colourful ornaments. They are flighty and will roost safely in trees or other high places. They do not scratch for food as chickens do, but rely on sight to spot insects and other pests; they are therefore less harmful if they get into a garden. They will create a great clamour whenever a stranger approaches. They will make a nest and sit their own eggs, but they are not such good mothers as chickens. During the day they are quite good at warding off predators, but they may fall to foxes when nesting on the ground at night. For purely domestic purposes, Guinea-fowl may be purchased in a variety of shades. Guinea-fowl eggs can be eaten like chicken eggs, and they have excellent qualities for baking meringues and the like.

Commercially, Guinea-fowl are reared only as table birds. Their meat has a more 'gamey' flavour than chicken's; it is esteemed in France. Commercial flocks can be run outside at quite high stocking densities. The modern, fast-growing commercial strains are mostly F 1 (first generation) hybrids, and they are best reared indoors. The newly-hatched birds, known as keets, may be sent by rail like day-old chicks, but they are often more expensive than the latter. It is not easy to acquire the correct parental strains for producing these F 1 hybrids yourself. Kept indoors, they are prone to panic; therefore corners should be netted off to prevent crushing, and lights should be operated by a dimmer switch to avoid sudden changes. Eggs should be hatched in incubators, and the young brooded for a time under heat. Growing birds should be reared inside at a moderate stocking density, to avoid a dusty atmosphere.

The demand for dressed Guinea-fowl is limited at the present time in the UK. It could increase in the future. Possible outlets are up-market restaurants, hotels, and game-butchers. We add a word of warning, for we have seen dressed Guinea-fowl imported from France at a price that would be hard to beat.

Turkeys.

Turkey breeding is a specialized business. The breeding, rearing, and fattening of turkeys is largely the province of a highly developed, intensive, turkey industry, in which processing, packaging, and marketing play as great a role as farming. Even in that industry, some of the smaller producers have experienced financial problems resulting from increased processing costs incurred to satisfy new hygiene regulations.

Opportunities for the smallholder lie largely in buying young, or growing turkeys and fattening them for the Christmas trade. The key to success for the smallholder is to sell dressed birds direct to the consumer, with the benefit of the retail price. To embark on this, you must first ensure that you can satisfy any regulations relating to the preparation, etc. of dressed turkeys. Second, you must find customers; this depends a great deal on your character. If you are affable and sociable, taking an active part in rural affairs and social events, chatting to people in the local pub, attending church or chapel, you may find plenty of people willing to order their Christmas turkey from you. If you are not sociably inclined, you will have to advertize, or seek a retail outlet, selling your birds at a lower profit.

Quail.

Gourmets will pay a very high price for quail eggs, and dressed birds can be sold to high class restaurants and game-butchers. Quail lend themselves to semi-intensive rearing conditions, and a small commercial breeding flock would not take up a great deal of out-building space. The main problem with quail is that it is easy to saturate the market; there is virtually no sale for surplus eggs or dressed birds outside the gourmet trade.

Incubation.

Incubation may be natural, or artificial.

Natural incubation. With the exception of domestic ducks, most poultry will incubate their own eggs, and raise their progeny. Broody hens are often used to hatch the eggs of other poultry, such as ducks; Silkie hens make good broodies. When a chicken goes broody, she should be moved to a broody coop where she will have protection from predators. The coop need not be large; a tent-shaped, or rectangular box, covered over at the nest end, and with wire mesh or slats at the other end, is adequate. The nest should be a piece of turf to provide some humidity, with straw or hay on top. Food and water should be made available near the nest. The hen may be placed on the nest, and as many fertile eggs as she can cover introduced beneath her, from behind. Check to ensure that she has settled down on the eggs, after which there should be little trouble. Do not mix eggs of different species, such as chicken and duck, for they will not hatch at the same time, and those requiring longer incubation may be abandoned. After hatching, the hen will look after the chicks, ducklings, etc., but they will need trans-ferring to a larger brooder. Geese and muscovies will sit their own eggs and look after their young. All members of a set of geese, the gander included, will take part in protecting the young. At this time geese will become quite aggressive, and defenceless poultry such as ducks should be kept away from them. Guinea-fowl, geese, and muscovies will all lay two clutches of eggs. If you have an in-cubator, it is a good idea to incubate the first clutch artificially, leaving them to sit and rear the second. This will probably increase your production. On removing the first clutch you may have to take action to break their broodiness by keeping them off their nests, out and about. One problem which sometimes arises with geese run in sets, is that all the females of a set may decide to sit together on a single clutch of eggs; provision of packing cases containing nests, one for each goose, sometimes helps to prevent this problem.

Artificial incubation. It is possible to make your own incubator, but to make a good and serviceable one is quite complicated. For any serious poultry enterprise it is best to purchase ready-made in-cubators. There are plenty of small electric incubators on the market. Basically, they fall into two types - still air machines, and forced draught machines. In the latter, warm air is circulated at the correct temperature by fans; in the former, the heat comes from above. Many of these machines require you to turn the eggs periodically, by hand; others have semi-automatic, or automatic turning devices. These incubators also contain a water reservoir, often with some means of adjusting the relative humidity. If you have no electricity, you will have to look round for a paraffin, or gas incubator.

Hen eggs hatch after three weeks' incubation; Guinea-fowl, duck, and turkey eggs take a month, goose eggs a few days more. Eggs of different sizes, such as goose and duck, should never be set to-gether in a still air incubator, for the heat coming from above would have to 'cook' the tops of the goose eggs for the duck eggs to be at the right temperature. Eggs of waterfowl should not be set with those of land birds; the former require relative humidities in excess of 60%, whereas the latter require only about 50%. It is also unwise to set eggs with different incubation periods at the the same time, for the first to hatch may contaminate the rest.

Before setting eggs, you should run your incubator for a few hours and check that it is at the correct temperature setting - 37.5°C for forced draught machines, and 39.4°C just above the eggs in still air machines. The thermostat may need adjusting to obtain the correct setting. Some modern incubators have electronic temperature control, and should not require adjusting. Eggs to be set in machines without a turning device should be marked by soft pencil with opposite noughts and crosses; this makes it easy to check that all have been turned. If you have the time, turn the eggs five times a day; three times a day is said to be the minimum, but some people get away with twice a day. Every day, the temperature and humidity should be checked; for the latter you may use a dial hygrometer, or the more accurate wet bulb thermometer. The difference in temperature between your ordinary thermometer, and the wet bulb one will give you the wet bulb depression; at 50% relative humidity it is 8.8°C, at 60% it is 6.8°C.

After a week, the eggs can be candled. An egg candler can easily be made, following the diagram in fig. 6-1. The light is switched on and the eggs are held against the hole in the candler. A fertile egg shows a dark streak, often with a network of veins visible around it; an infertile egg appears clear. The latter may be hard-boiled, and served chopped, to young birds.

You should stop turning the eggs when they start to 'pip.' When hatching begins, avoid the temptation to help them out unless they are really stuck - helping them can cause more harm than good. The newly-hatched birds should be left in the incubator to dry off and 'fluff-up.' Then they may be transferred to a warm brooder, and the incubator cleaned out. If any infection is suspected in the incubator, it may be fumigated by placing a non-metal, and non-corrosive container of formalin inside, adding a little potassium permanganate, and closing the lid. This operation is best done outdoors.

Brooding.

After hatching, young birds should be warm-brooded for a time, if possible with a gradually reducing temperature regime to harden them off ready for transfer to an unheated follow-on brooder.

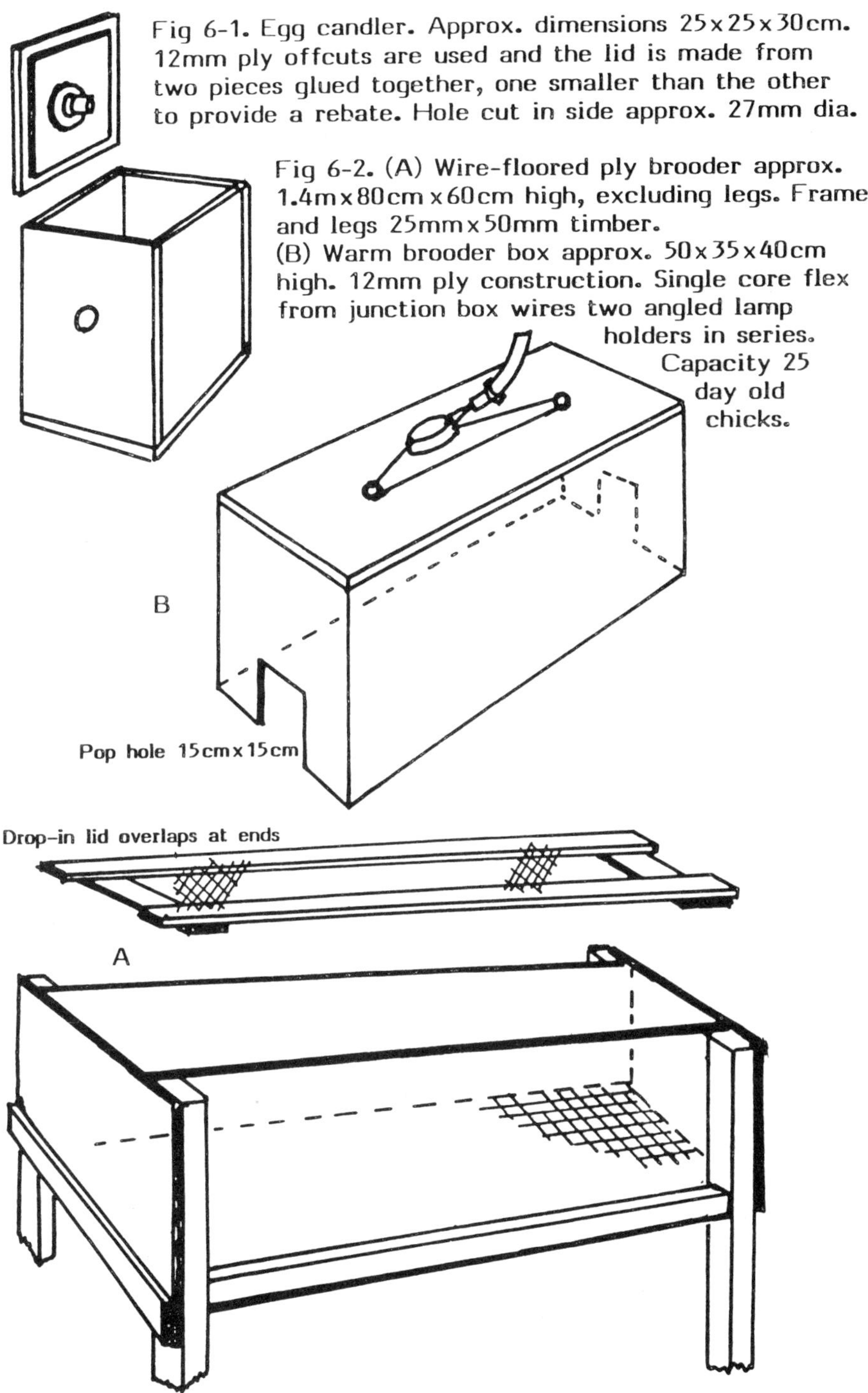

Fig 6-1. Egg candler. Approx. dimensions 25 x 25 x 30 cm. 12mm ply offcuts are used and the lid is made from two pieces glued together, one smaller than the other to provide a rebate. Hole cut in side approx. 27mm dia.

Fig 6-2. (A) Wire-floored ply brooder approx. 1.4m x 80cm x 60cm high, excluding legs. Frame and legs 25mm x 50mm timber.
(B) Warm brooder box approx. 50 x 35 x 40 cm high. 12mm ply construction. Single core flex from junction box wires two angled lamp holders in series. Capacity 25 day old chicks.

As a rough guide, land birds need about four weeks' warm brooding; waterfowl about half that time. Warm brooding should be continued for a longer period in cold weather.

A design for an electric brooder is shown in fig. 6-2 A. This is simply a rectangular plywood box, with a welded wire mesh floor, and a framed lid covered in wire netting. The outside of the box may also be sheathed in netting, or thin metal sheeting, as a protection against rats. Heat can be supplied from an infra-red lamp, sitting initially on top of the wire netting lid. The temperature can be lowered progressively by raising the lamp a little at a time, or by placing a dimmer switch in the circuit. A 250 watt infra-red lamp will use 42 units of electricity a week, so it may be worth investing in a dimmer switch which will reduce the consumption of electricity as less heat is needed. An economical alternative form of electric heater, shown in fig. 6-2 B, is made from a bottomless plywood box with two pop holes at opposite ends. Two angled lamp holders are fitted inside the 'lid,' wired in series. A junction box is fitted on top of the box, and a single core flex runs to one of the lamp holders, another single core flex runs from that lamp holder to the second, and a third flex runs from the second lamp holder back to the junction box. Two 150 watt bulbs are fitted inside; wired this way they consume about 75 watts between them, and they give out less visible light and more infra-red. Wired this way, they also last longer. Progressive reduction in temperature may be achieved by replacing the pair of bulbs with two 100 watt, then two 75 watt bulbs, or by fitting a dimmer switch in the supply circuit. In cold weather, it may be necessary to start with a pair of 200 watt bulbs. We have used this design many times and have experienced no problems from young birds pecking at the light bulbs nor from over-heating of the top of the plywood box. If you are worried about bulbs breaking, or fire risk, some soft wire netting may be shaped as a push-fit to protect the light bulbs; the exterior box may be made from thin metal, such as aluminium, which you can cut with tin snippers. With a metal case, be sure that the bases of junction box and lamp holders are separated from the metal by insulating ma-terial, and that the metal casing is properly earthed. This type of unit can be placed inside the larger wire-floored brooder. There are ready-made units of this type for sale, such as the 'Ali' brooder.

Before introducing newly-hatched birds into a heated brooder, cover the wire mesh floor with two thicknesses of newspaper, and switch on the heating system. Food and water must be provided, initially as close as possible to the warm brooding area. Ensure that the food and water containers are sufficiently illuminated for the birds to see them. A first feed of chopped hard-boiled egg is good, followed by proprietary brands of starter crumbs. For land birds,

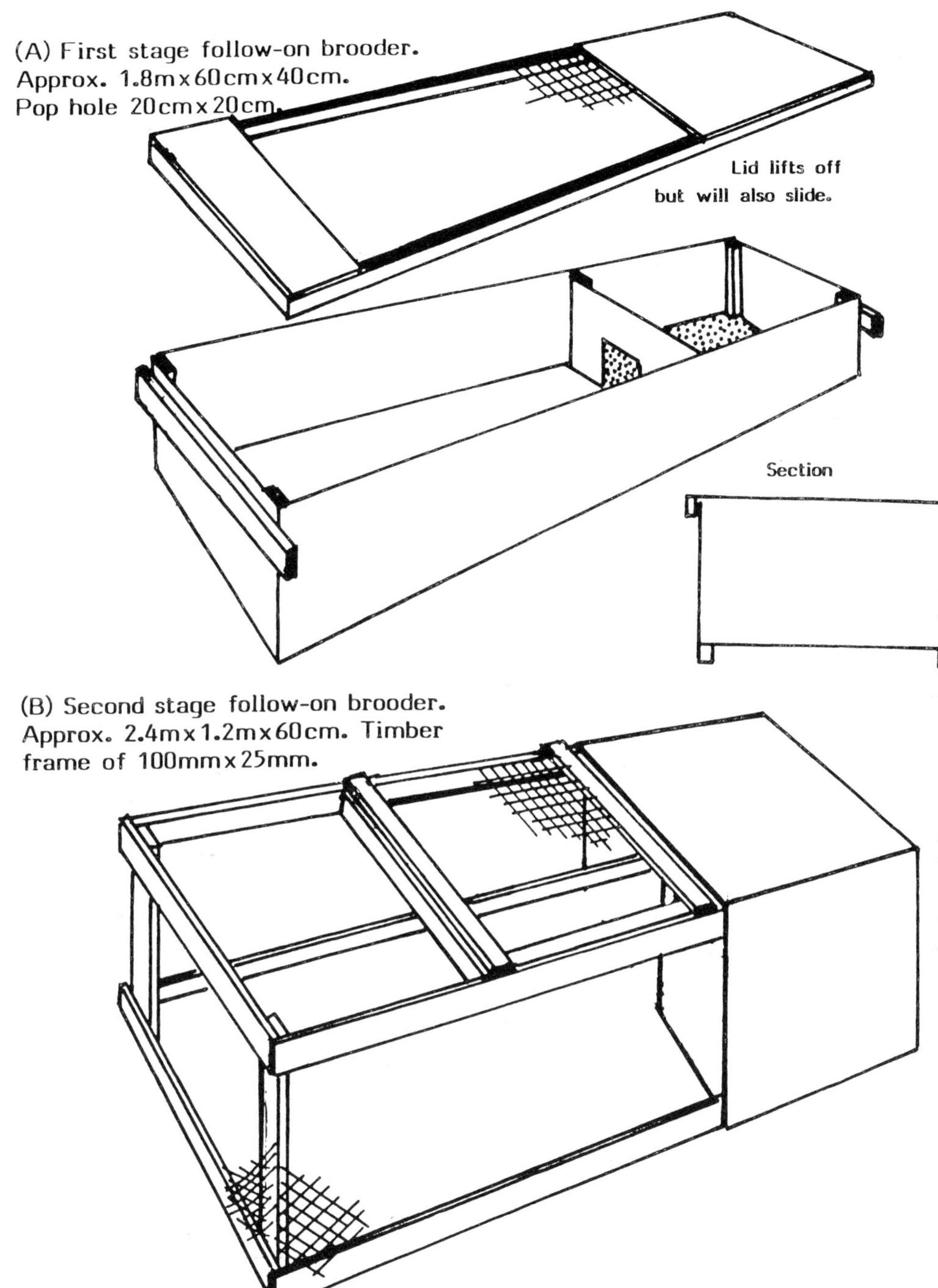

Fig. 6-3. Follow-on brooders.

water may be in a shallow container, to avoid any risk of drowning; ducklings and goslings must be able to immerse their heads in water, so that a deeper container has to be used - we use ramekin dishes for the first few days. The young birds may now be introduced. If you are using an infra-red lamp, the young birds will tell you if the temperature is too high by leaving a clear space directly under the source of heat; if the temperature is too low they will huddle together in the centre. If you use a heater unit with two light bulbs, you can check the temperature by placing a spirit thermometer on the floor inside; it should be around 35-37°C at the beginning for land birds, and about 32°C for waterfowl.

From then on, brooding should be straightforward, involving replenishment of food and water, and checks that all the birds are alive and well. Ducklings and goslings splash a great deal when drinking, and their food and water should, after the first few days, be moved as far away as possible from the warm brooding area. The newspaper around their drinking and feeding area will soon disintegrate and subsequent splashings will fall through the wire floor, where it may be absorbed on a layer of peat or sawdust. At the end of the warm brooding period, the young birds may be transferred to a follow-on brooder.

If you have no electricity, there are gas brooders on the market. Alternatively, small numbers of birds may be raised in a hay box brooder, which relies on conserving the heat generated by the young birds themselves.

Follow-on brooders.

A design for a first stage follow-on brooder is shown in fig. 6-3. Basically this is a long, rectangular, bottomless box made from exterior ply or oil-tempered hardboard. One end contains a covered nesting area, entered through a pop-hole; the other end is covered to protect the food from rain. The rest is covered in wire netting. We find these brooders very successful outside on short grass - the birds soon clear up the grass, and the brooder should be moved over fresh grass as necessary. The floor inside the nesting area should be covered in straw, peatmoss, or wood shavings. As the birds grow they will need transferring to second stage brooders, such as the mobile arks shown in fig 6-3, or slatted-floor arks kept in pens. By this stage, chickens not on slatted floors should be provided with perches. Throughout these stages of brooding, your birds may be fed on the appropriate growers' mash or pellets.

With the exception of grazing birds, such as geese and muscovies, it is not essential to brood outdoors in the way we have described.

If you are keeping your birds partly, or entirely indoors, in straw yards or large sheds, you may also brood indoors. Suitable brooding areas may simply be partitioned off, using any kind of sheeting, or concrete blocks, and heat may be supplied from an infra-red lamp, or one of the heating units described above.

Sexual differentiation.

As your birds grow they will begin to show signs of sexual differentiation. In many breeds of chickens the pullets' combs are smaller than the cockerels'; the pullets' tails are pointed, whereas the cockerels' have more small feathers which become curved as they grow. In coloured breeds of ducks, the drakes become more strongly marked and colourful than the ducks, especially on the head and neck. In all breeds of ducks the drakes develop curled tail feathers, whereas the females' are straight; drakes cannot make the distinct quacking sound of the female. Likewise female Guinea-fowl are more vociferous than the male. Sex differentiation in geese is more difficult. Behaviour sometimes helps, for the males will go in front and adopt a threatening posture when a stranger approaches. Adult geese may be sexed by opening the vent; in the female the inner surface is loose and wrinkled, dark pink; in the gander it is smooth and pale pink, and the penis is white. Many goose keepers are, in fact, able to sex their geese quite accurately, without being able to explain clearly how they do it.

Housing.

Some smallholders may wish to keep only a few hens to provide themselves with fresh eggs. The mobile ark shown in fig. 6-4 is quite suitable for this purpose. It is designed to be made from two-and-a-half sheets of 6mm exterior ply, a quantity of 50mm x 25mm timber, and some chicken netting. It can house four or five laying hens and it may be used as a mobile ark, moved as the grass is eaten up, or it may be provided with a pop-hole to let the hens run outside. A slatted floor is shown under the covered section, but this may be dispensed with and two perches fitted instead. An outside nest box may be added to increase the accommodation slightly. The movable top of the ark must be securely fixed by hooks and eyes, or by heavy weights, or it will blow off in strong winds. One of us started off in poultry with one of these arks, and found it quite successful, moved progressively around the kitchen garden and lawn. The ark shown measures about 2.4m x 1.2m; the design may be enlarged to, say, 3m x 1.5m, to house seven or eight laying hens.

General purpose houses, which are light, airy, and draught-free, for birds kept in pens or on free range, may be constructed using

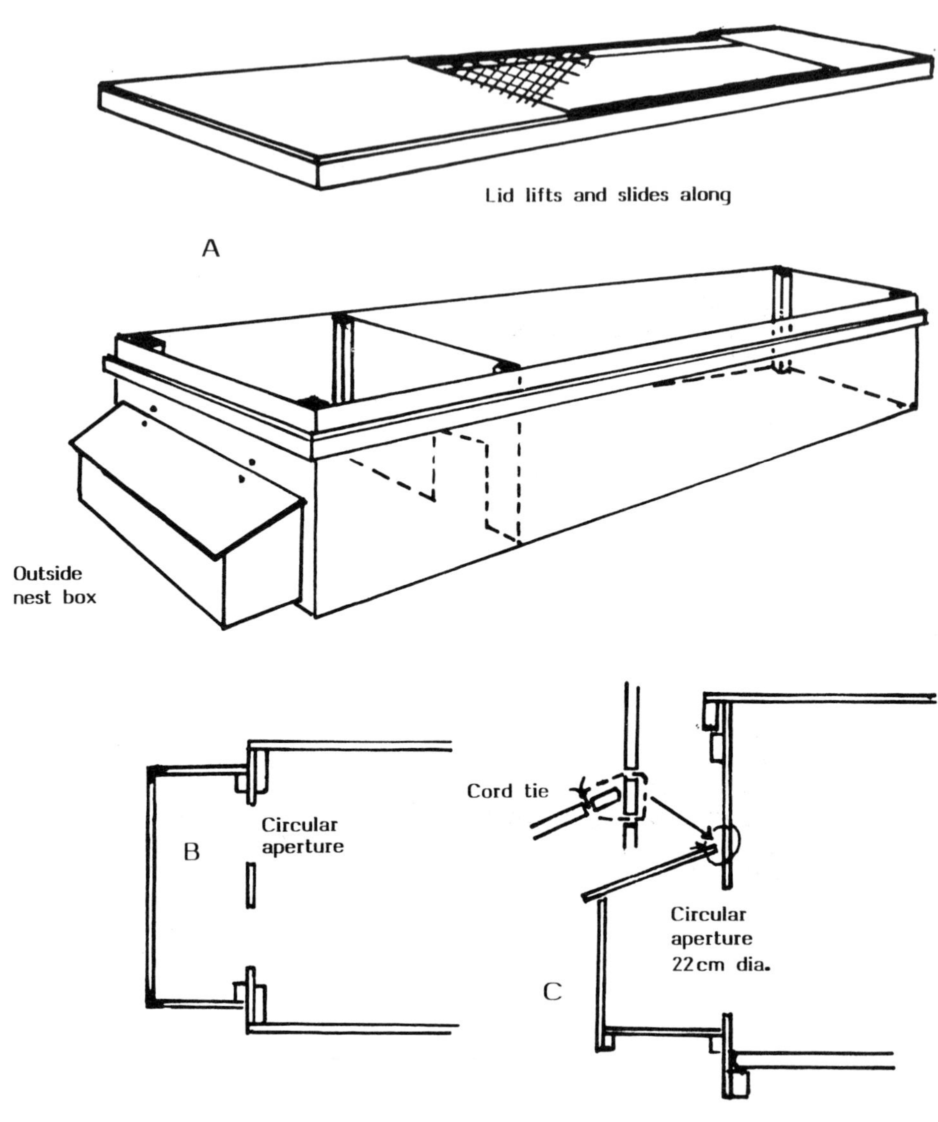

Fig. 6-4. Mobile ark. (A) Ark and lid. (B) Horizontal section through nest box. (C) Vertical section through nest box.

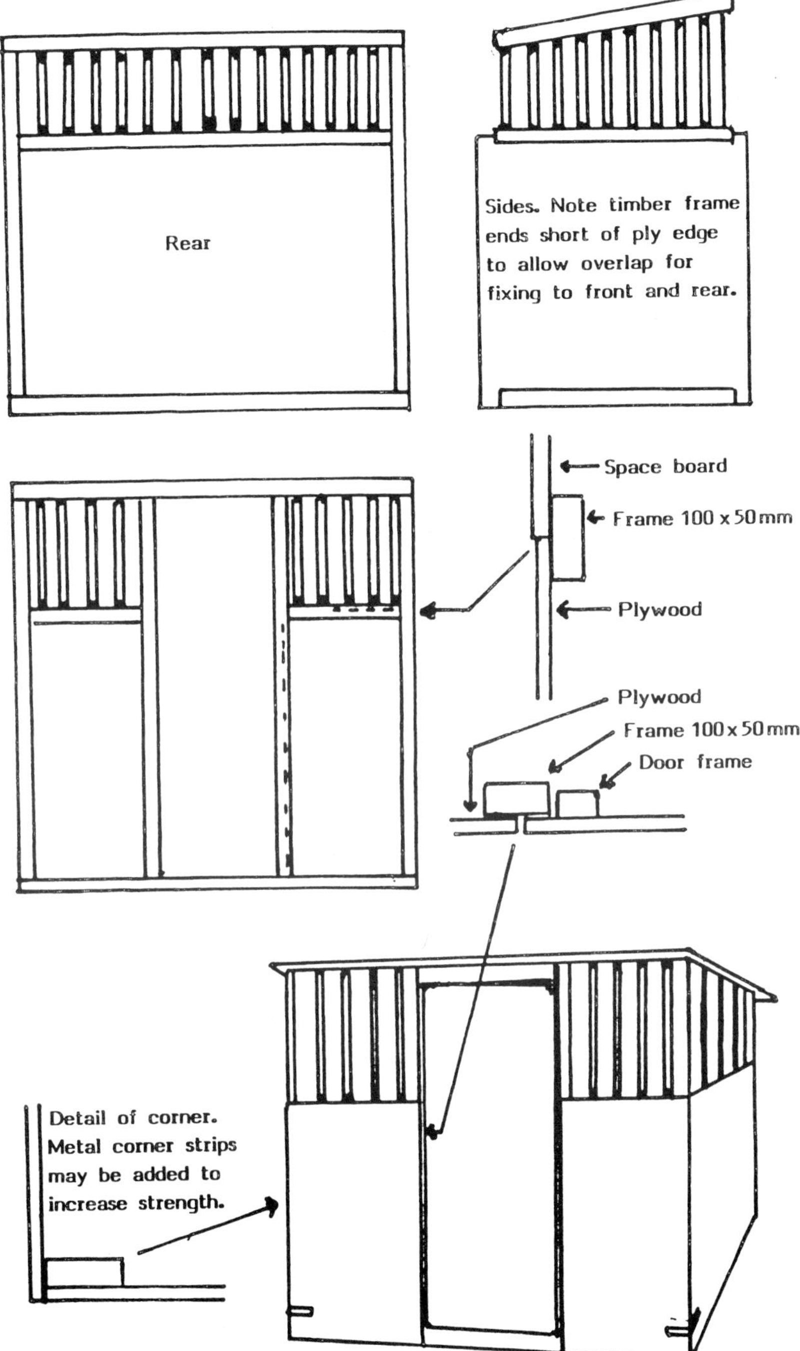

Fig. 6-5. Poultry shed.

87

6mm to 12mm exterior ply (or 12mm shuttering ply), 75mm x 50mm sawn timber (such as secondhand floor joists), and cheap, short offcuts of approximately 100mm x 18mm sawn timber from a sawmill. The roof may be of ply, or of corrugated sheeting. For economy, houses should be designed to avoid waste of the plywood sheets, which generally measure 2.4m x 1.2m.

Fig. 6-5 gives details of a house measuring about 1.8m x 1.2m. The front, back, and sides can be prefabricated separately on the floor, or a large bench; this makes the construction much easier. The two sides are then fastened to the front and back, and a roof nailed in place, overhanging a little all round. A solid timber, plywood, or slatted floor may be fitted, in sections which are removable for cleaning. A few holes should be drilled in solid floors for drainage. Alternatively, the house may be left without a floor, with peatmoss, wood shavings, or other litter covering the ground, the house being moved when it needs cleaning. Personally we find floorless houses suitable for ducks and geese, and slatted floors useful for chickens. If chickens are kept in a house without a slatted floor, horizontal perches of 50mm x 50mm timber, rounded on the top edges, should be fitted. Always fix the perches at the same height; this avoids disputes over the highest perch. The ends of the perches should slide into square slots to permit their removal for cleaning and anti-mite treatment. For laying hens, it is best to fit this type of house with an outside nesting box, access being via circular holes cut into the back of the house. In exposed situations, small houses can sometimes be blown over by the wind. Transverse runners can be bolted across the bottom, projecting outwards some distance, to prevent this. A house of this size would be suitable for a set of geese, eight to ten ducks, or a dozen laying hens; it may be sited in a pen, or used on free range.

Using the same construction system, quite large poultry houses can be prefabricated in sections, each based on a single sheet of ply. The sections can be set up on a concrete or hard core base and bolted together. The houses may be designed narrow, with a lean-to roof, or wide, with an apex roof. The advantage of prefabricated, bolted sections is flexibility; you can modify, move, or sell your sheds if your farming policy changes. The construction of such sections is shown in fig. 6-6; only the basic idea is shown, without details of roof construction, which may be easily worked out. Remember that the roof timbers must be attached to the uprights with metal roof ties, and the uprights themselves should be bolted to posts sunk in the ground, or to metal ties passing into concrete set in holes in the ground. The larger the building, the greater is the effect of wind pressure. A cheaper alternative for large poultry sheds of this kind is to make up a timber framework, anchored

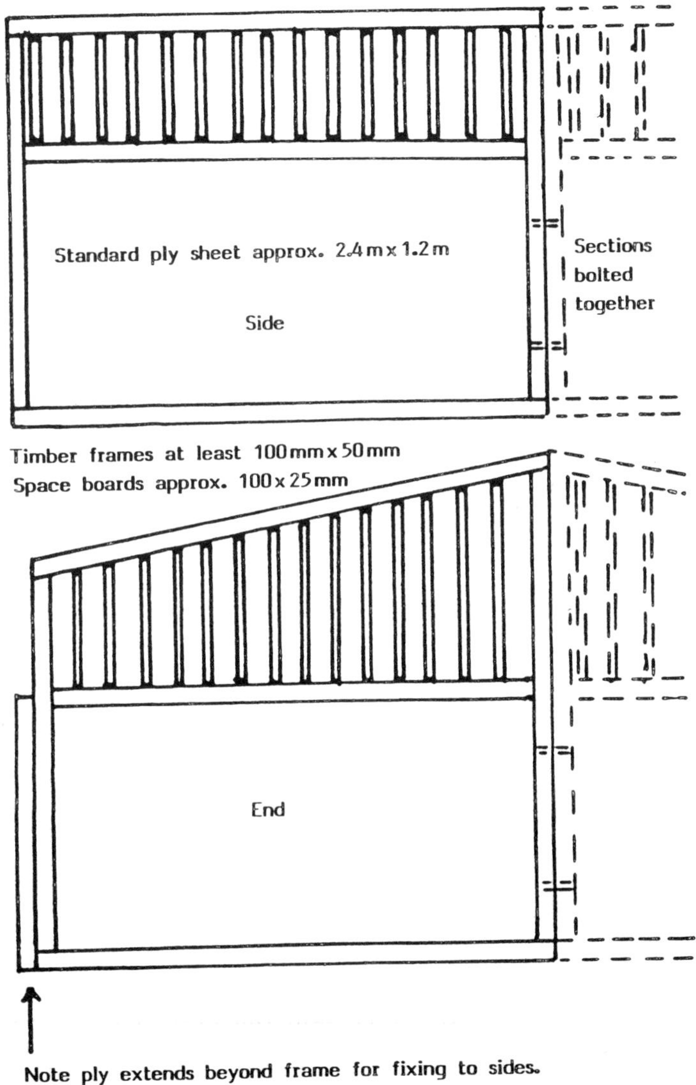

Fig. 6-6. Basic construction of sections for larger sheds.

in the ground, with used conveyor belting fastened round the outside, up to the height of the space boarding. - Larger houses of
this kind are suitable for the straw yard system of semi-intensive
poultry keeping. In these systems the straw litter is thoroughly
cleaned out at much more frequent intervals than in deep litter
systems. The house may have an aviary-style outside run added,
covered in wire netting. The run may be roofed over, but the
uprights and roof attachments must be secure enough to prevent
the wind from lifting off the roof. The advantage of a roof is that
the straw litter lasts longer if protected from the rain, and wild
birds can be denied access to the poultry food. If the run is not
provided with a solid roof, it may be protected from wild birds
by a covering of fine netting; in winter this must be replaced by
wide mesh 'pigeon netting' which does not accumulate snow.
Alternatively, if enough land is available, the straw yard house may
be given access to an outside grass run, used only for a few hours
at a time in dry weather, to avoid poaching. Straw-yard housing
is not economical for grazing birds such as geese and muscovies.
It is a practical method of keeping laying hens under semi-intensive
conditions, and for guinea-fowl; breeding flocks of ducks may also
be kept in this way, but it may deny them access to a pond, which
some people think is cruel. Straw yards are also suitable for
rearing and fattening table birds; chickens, guinea-fowl, turkeys,
and ducks.

Feeding and general management.

Free-range birds. The term 'free range' is perhaps best defined in
terms of stocking density, for poultry kept in a large field with
poultry-proof fences are not free to go elsewhere. A stocking rate
of a little over 100 hens per acre, or not more than 250 to the
hectare, should qualify as free range. In our opinion, therefore, ten
or fifteen hens kept in a pen of about one tenth of an acre should
also qualify as free range. Naturally the stocking rate would be
greater for smaller birds, such as guinea-fowl, and considerably
smaller for large birds such as geese.

Free-range birds generally do well on a diet of grain alone. Some of
the mixed corn on the market is of doubtful quality - we have
found flaked barley to be one of the best feeds for free-range hens,
ducks, and geese. Feeding is most economical if you have rat-proof
hoppers which can be opened, morning and night, for a fixed time;
otherwise it is best to give only as much as they can clear up in
a reasonable time, at each feed. Chickens must be fed in daylight,
but ducks and geese will feed quite well after dark, when they have
been shut in for the night. Grazing birds, of course, only need
feeding in autumn and winter; other birds need feeding throughout

the year, but they will obtain part of their food in spring and summer from grass, clover, seeds, insects, and the like. Water must always be available, and ducks and geese must have it deep enough to immerse their heads. On free range, your birds may find grit which is essential to the proper functioning of their gizzards. It is still advisable to provide a grit box; some fine gravel should be made available for geese. Chickens enjoy a dust bath; they will often find their own places on free range, but a box of fine earth may be provided, protected in some way from the rain.

Household scraps may also be fed to poultry, boiled and served as a mash. You should avoid using very salty scraps - they should not have too much salt.

The houses of free-range birds should be cleaned out when needed. The ends of chicken perches, and nooks and crannies, should be treated with liquid paraffin, or similar, to kill mites. Some poultry keepers periodically treat their houses with creosote or Stockholm tar; if you do this, no birds should be returned until all trace of smell has disappeared and the wood is quite dried out.

We have already pointed out the importance of good, short grass for grazing birds; we would also add that long grass should be avoided for all poultry because it provides better conditions for the survival and transmission of their parasites.

As a protection from predators, free-range birds should be shut in at night. Chickens will 'put themselves to bed' when it begins to go dark. Ducks and geese are naturally night feeders; as a rule they have to be driven into their houses. To drive these birds, wave your hands in the air above them. If you give them an evening feed inside their house, it may encourage them to go in. Sometimes they behave like naughty children, repeatedly running past their door; a hurdle placed at an angle to their door will often help to direct them in. Water should always be available to birds shut in their houses. Rats can become a problem, often living under the houses in winter; we find poisoned food the best solution: A wooden box or crate is placed, inverted, on the ground with a couple of rat holes cut in the sides. Poison bait is placed inside, at the centre, well away from the entrances. Some people add a short length of ceramic drain pipe to each entrance, to protect small domestic pets from going inside. If the box or crate used is not heavy, a concrete block, or similar, may be placed on top as a safety precaution. We have found this to be the best way of dealing with rats; if you leave them alone they will steal a great deal of food, and kill your young birds.

Semi-intensive poultry systems. On a small scale, birds may be kept in small houses, each with its own pen, which should be provided with straw litter. On a larger scale, a straw yard system may be a fixed combination of house and run. All birds kept under semi-intensive conditions should be given the appropriate proprietary foods. Grain alone may be insufficient when they do not have access to free range. Feeding frequencies, and quantities, will depend on the species, and whether breeding, growing, or fattening birds are being kept. Grit must be provided in boxes; also ground or flaked oyster shell for laying birds. Dust baths should also be made available, especially for chickens. Water must be available at all times; in large units it is an advantage to install an automatic watering system. In these semi-intensive systems, periodic cleaning out is essential; otherwise coccidial, and other disease, may occur. Ideally, if several straw yard units are set up, they may be used on a rota, one of them being kept empty for a time whilst parasites die off, or thorough cleaning and disinfecting is carried out.

We do not consider fully intensive systems to be within the scope of this book; they belong to farming as an industry, rather than as a way of life.

CHAPTER 7. OTHER LIVESTOCK.

Goats.

Goats are browsing, rather than grazing animals; if let loose in a field, they will prefer the hedgerow to the pasture. They can be used to graze marginal scrub land that is unsuitable for other stock. Goats originate from arid climes; they are more at home in Mediterranean countries than in Britain. They can withstand cold, but their hair has poor waterproofing properties, and they need shelter in prolonged wet weather. Feral goats have, however, survived outdoors in the British Isles, on the Burren of Clare in Ireland, and, until some years ago, at Malham Cove in Yorkshire.

Goats are not easy animals to manage; they need a lot of supervision and attention. In Mediterranean countries they are run in very large herds, under the supervision of a full-time goatherd. Most goat keepers in Britain keep only a few animals; this can mean that they take up a disproportionate amount of time in relation to the return they give.

Although the meat of young goats is of good quality, it has no commercial value in Britain. Goats are kept here mainly for supplying domestic milk, which some keepers make into a soft cheese. They are perfectly good for this purpose, provided you are prepared to take all the trouble of looking after them.

At the present time, almost the only people who make money out of goats are pedigree breeders who sell milking nannies to other farmers, smallholders, and backyarders. A few years ago, a goat's milk venture was tried on a commercial scale at Tregaron in West Wales. The idea was to sell goat's milk for consumption by those people who are unable to digest cow's milk. The milk was powdered to make for easy storage and transport, with a view to distributing it to hospitals and retail outlets throughout Western Europe. The cost of processing relatively small quantities of goat's milk to powder, made the product seem expensive, and the project did not succeed as well as had been anticipated. Goats do, however, have some commercial possibilities for enterprising development. Unlike cow's milk, goat's can be frozen; in this state it could perhaps be successfully marketed in a more restricted geographical area at a more reasonable price than a powdered product. The French produce a wide variety of 'fromages de chèvre,' some of which are excellent and sell at a high price; they also produce very good 'pâté de chèvre.' Someone who could find out how to make cheeses and pâté to the French standards could make goat keeping commercially viable in Britain. Good quality goat's cheese could probably be promoted in the UK; goat pâté would be more difficult - it might sell at home as 'French-style country pâté' with the goat's meat ingredient in small print, or it may have to be exported to the Continent. Young goat's meat, suitably packaged, could also have export potential to the more southern parts of Europe. Such a business venture would, of course, involve capital risks; it might be worth trying if support were available from a government-backed Development Agency, or similar.

There are three different ways of keeping goats:

1. To allow them to graze pasture. You will require extra-high fences. Goats will climb over fences erected for sheep or cattle; in so doing, the nannies often tear their udders on the top strand of barbed wire. A simple shed, in which they can retreat in bad weather, is advisable.

2. To keep them securely tethered on their feeding ground. Using this system, you can make your goats graze and browse those parts of your land that are least suitable for other stock. Tethered goats need a fair amount of supervision, for they often become entangled in their tethers; some people think tethering is cruel. Many people who tether their goats, bring them in at night.

3. To keep them indoors, with access to a high-fenced concrete yard, and perhaps the occasional luxury of supervised feeding out-side. As with other stock, the buildings need to be ventilated, but draught-proof. Inside pens and the yard should be designed to facilitate cleaning out. This is a labour-intensive method.

If you wish to obtain a high milk yield from your nannies, you should buy stock from a herd with a good milking pedigree, and feed them adequately. Just like milking cows, concentrate feed is needed to obtain maximum milk yield; otherwise you will have to be content with just a small yield.

New commercial developments apart, our opinion is that goats should be kept for domestic milk only by those smallholders who have a preference for them, and are also in a position to devote sufficient time to them. They will mostly be people who have some source of private income; those who intend to rely on a holding as their sole source of income would probably be better off with a good house cow which will give them milk, and a valuable beef calf. Those who keep goats could be doing a service for the future, for, should the wheel of fortune turn and oblige more and more people to return to a subsistence economy, the goat would then assume real importance.

Horses.

The breeding, rearing, and breaking of horses is a specialized business requiring long-term investment, for they are slow-maturing animals. Unless you already have considerable experience and expertise, you should not even think of such a venture. Converting a smallholding to riding stables, or a pony trekking centre, requires not only experience with horses, but also considerable business acumen, and the holding to be sited in an appropriate geographical location. If you are seriously thinking of such a venture, you will not need the sort of basic information we are giving in this book.

If you are fond of riding, you may, of course, keep a horse for your own use, if you can afford this luxury. A horse needs a good two acres of grazing. Cattle are needed to follow-on the grazing, to clear it up and keep the land 'sweet.' If you buy a sturdy horse, it may also do some work for you, such as dragging a sledge of hay bales in winter when the ground is unsuitable for a tractor. A working horse can be a practical proposition on the land, but in practice, very few people successfully manage their holdings in this way. Beware of buying a working horse just as a romantic whim, thinking you will use it to go shopping, and work the land - in practice, as soon as the novelty wears off, you will find yourself jumping into your car to go shopping, as a matter of speed and convenience, or using a tractor to get a job done quickly on the land.

There are, however, possibilities of making money from horses on a smallholding. If you have some surplus grass in the autumn, you may let out the land on tack for horses from, say, a pony trekking centre. This can be quite remunerative, and the ponies will benefit not only from the food and a good rest, but also from being on pasture free from horse diseases. However, they will crop your grass very short, poach the land in wet weather, and loosen your fencing stakes by feeding over the other side, where 'the grass is greener.' Tack horses can be useful for clearing rough land - they may leave behind a 'moonscape,' but it will be in a better state for carrying out grassland improvement. - If your holding is in a suitable location and some of your land has road frontage, you could consider fencing off one or two small paddocks, providing each with gated road access, water, and a simple shed. In the right areas the rent from such horse paddocks could be much higher than the return you could expect from farming the same area of land. To safeguard yourself from establishing a legal tenancy, you would have to enter into letting agreements not exceeding 364 days, with the land closed off for at least one day a year. You should, of course, seek legal advice before making such agreements.

Pigs.

Pigs are rooting, rather than grazing animals. Thus their only relationship to grassland farming is that they may be used to clear rough land, prior to re-seeding. In years gone by, many farmers and smallholders used to keep a breeding sow, or rear and fatten a bought-in piglet. Nowadays it is uneconomical to fatten a pig for the freezer if you have to buy commercial feed in small quantities. It is only feasible for those who have a lot of orchard windfalls, or for large families producing large amounts of vegetable scraps or having surplus milk from a house cow. If you do keep a pig, all waste food should be cooked for it, and no meat should be included in the scraps; it is illegal to feed any meat whatsoever. If you do keep a pig, make sure its sty is strongly built; a pig can easily demolish an old stone sty if the mortar joints have weakened. Some people think it not right to keep a single pig, for it will become lonely; there is also the risk of it becoming a family pet, posing a problem when it is ready for killing.

Modern pig farming is based on specialized intensive factory-farming systems. Pig farmers buy their feed in bulk at a discounted price; some pig farmers form syndicates so that they can buy their feed at even bigger discounts. On a small scale, you would probably not be able to purchase pig food at a price that would make it economical to breed, or rear them.

Deer.

Venison can fetch quite a good price, especially if marketed direct to good class hotels and restaurants. Deer can thrive on marginal land which is not suitable for cattle. They will browse on young trees and shrubs; they can therefore be kept on scrub land and in open woodland where there is some ground vegetation and young coppice (which they will eat). The biggest problem with deer is the high cost of fencing needed to keep them in - for red deer the perimeter fence needs to be 1.8m high. You should not think of keeping deer unless you are buying the right kind of land, at a price per acre that will compensate for the high cost of perimeter fencing. Furthermore, for fencing costs to be economical, the area of land given over to deer would need to be fairly large, and about as long as broad. A high fence round a small area of land, or round a long narrow strip, would represent too high a cost per acre to be economical. Another point to bear in mind is that deer-farming is still in its infancy, and the disease risks of keeping them in higher population densities than occur in nature have not yet been fully assessed. So far, in the United Kingdom, only a few people have managed to establish successful deer enterprises.

Rabbits.

The famed prolificity of rabbits gives them appeal to the small-holder interested in semi-intensive meat production. As grazing animals they can be kept in wire-floored arks on grass during the growing season. Keeping a large number of rabbits this way entails quite a lot of labour, for the arks have to be moved frequently to fresh grass. There are rabbit specialists about the country who will set you up with breeding stock, and all necessary cages and equipment. If you decide to try rabbits, buy your stock, if possible, from a specialist breeder-dealer who will guarantee to market the rabbits you produce. Some of the people we know who have tried rabbits, have not succeeded. They have ended up spending more on rabbit feed than they were receiving from meat sales. This financial deficit often seems to have resulted from losses of very young rabbits, just after birth, making the meat output small in relation to the cost of feeding the adult breeding stock. For success, a great deal of time needs to be spent in supervising the rabbits, especially when they are near to giving birth, so as to avoid losses at this time and ensure that the young are reared success-fully. The only people we know, or know of, who have made a success out of rabbit breeding, do it as a whole-time occupation, on a fairly intensive basis. We think, therefore, that rabbits should be considered only by those who are sure they will be able to devote a major part of their time to them.

CHAPTER 8. OTHER SOURCES OF INCOME.

The majority of grassland smallholdings in Britain are not viable when run solely for livestock rearing. That is to say, they do not produce sufficient income to support a family. The only people we know who make a living from stock on a small acreage are horse, cattle, and sheep dealers. They have been born in the business and know it inside out; they spend a great deal of time at auction marts and run their holdings like retail shops, with a regular turn over of stock.

Those who have a small private income, and who only require their holding to produce some of their food, and a small cash surplus, may manage quite well. Those who have no private source of income, must seek to supplement their income from their small-holding. This may be achieved by seeking employment on a part-time basis, or by remaining self-employed and developing some home based activity in addition to extensive livestock rearing. Possibilities include specialized semi-intensive development of part of the holding itself, selling home-made goods, offering a service outside the holding, or developing some tourist potential on the holding itself. We offer some suggestions, as food for thought, under separate headings:

Flowers, fruit, vegetables, and Christmas trees.

You may wish to grow as much of your own fruit and as many vegetables as possible, preserving, storing, or freezing sufficient to last you through the winter. You may also grow vegetables such as curley kale, and purple sprouting broccoli, to have fresh 'greens' out of season. If you simply extend your vegetable garden with the idea of selling a little of everything you produce, you may find that your neighbours have surplus vegetables at the same time, and that the fuel costs of taking small quantities of this and that to town or market will reduce your profit. This procedure is only feasible if you have the possibility of farm-gate selling. If you extend your kitchen garden further, until it reaches the proportions of a full-scale market garden, it will become a full-time activity, in conflict with your stock-farming, especially at busy times such as lambing, calving, and hay making. If you want to be a market gardener, all well and good, but it should be your main activity. If you intend to run a holding as a livestock enterprise, you would be better off thinking in terms of a single cash crop which will take up relatively less of your time than would a variety of crops. Whatever crop you choose, you should first ensure that you will be able to market it, and that your soil, and the local climate, are suitable. Here are a few suggestions. There are many other possibilities.

Flowers. You could grow flowers on a regular basis, selling through retail outlets, or to the public outside the nearest hospital, if it seems feasible. You would need an unheated, and a heated greenhouse, to keep up a regular supply. A heated greenhouse would be expensive to run unless you used timber fuel cut on your land. An alternative would be to produce flowers only for a specific time, such a Christmas, growing chrysanthemums in an unheated polythene greenhouse. The latter could be used for some other crop, earlier in the year.

Strawberries. This valuable, but perishable cash crop needs to be sold quickly, by farm-gate selling, through retail outlets, or by self-pick. (The latter may not be successful if you only have a single crop, because of the cost of advertising). There are different varieties of strawberries which permit an extended fruiting season. Before planting a large area of strawberries, make sure that you will be able to market them profitably. You may have to await a second season before you obtain a good yield. Before planting, ensure that all perennial weeds have been eradicated. A light, well-drained soil is recommended. Old dung, compost, or leaf mould should be spread. Before fruiting, a mulch of straw, or similar, is needed to protect the fruits from touching the ground.

Soft fruits. By choosing the right varieties of strawberries, loganberries, raspberries, and perhaps black currants and gooseberries, you may be able to produce soft fruits over an extended season, your earliest fruits coming perhaps from an unheated polythene greenhouse. Assuming there were enough good class hotels and restaurants within easy driving distance, you could arrange to keep them supplied with fresh soft fruits for speciality deserts throughout the tourist season. Alternatively, you could go for self-pick, a long fruiting season making advertising worth while. In the latter case you would have to provide parking facilities, and perhaps a play area for children. Outside catering for afternoon teas could go with this activity. If you made a success of soft fruits, you would have to concentrate on this and carry out only the simplest forms of livestock rearing on the rest of your holding. One of the biggest problems with soft fruits is losses to wild birds; fruit cages will protect them, but they involve a high capital outlay.

Blueberries. Unlike the small British bilberry, the American blueberry is large and has commercial possibilities. If you have acid, peaty land, with underlying water, you could try this crop. It should be possible to promote and market it through retail outlets, without much difficulty. This would, however, be pioneering work, and we would advise you first to try only a few plants to see if they grow and fruit successfully on your land. If possible, choose a variety that does not need a pollinator. If successful, then, and only then, should you start growing on a commercial scale. Given success, you may end up selling the know-how, and the cuttings, to other small-holders. This would be a high-risk undertaking at the beginning; we mention it only because it offers possibilities for producing fruit on acid, peaty soil, where the water table is barely half a metre below ground, and which is unsuitable for most other crops. You should, of course, research this subject thoroughly, and seek additional expert advice, before embarking on such a venture.

Asparagus. A valuable cash crop can be produced from a set of asparagus beds. The price you should be able to obtain may make it worth transporting your produce some distance. It has an advantage over soft fruit in that it does not perish quite so quickly. Before establishing asparagus beds, all perennial weeds must be eradicated using a systemic weed killer. Light, sandy soil is best, but modern varieties will grow on well-drained clay soils. Cropping is later, on clay soils, because they do not warm up so quickly in the spring. Compost should be dug in the beds; in subsequent years it should be spread as a top dressing. Establishing asparagus beds is a slow process; if you plant out year-old crowns, it will take two or three years before you will have a saleable crop, and this means so many years of patient weeding with no immediate return. Once estab-lished, a bed may last up to twenty years.

Potatoes. A high-yielding crop, the potato could, in theory, close the financial gap on many smallholdings. Maincrop potatoes, grown under normal conditions and with the correct amounts of fertilizer, can yield 10 to 20 tonnes per acre. Seed potatoes can be grown with a view to obtaining a higher price for the crop; likewise, in the warmer south-western corner of Britain, early potatoes can be grown successfully. - There are snags, however. If you grow more than four-tenths of a hectare and intend to sell your crop, you must register with the Potato Marketing Board who will give you a 'basic area.' In any year which the Board designate as a quota year, you will be given a quota as a percentage of your basic area. If you wish to grow seed potatoes for sale, there are additional regulations to be complied with, and for classified seed potatoes your land must be officially examined to ensure that it is free of potato cyst nematode.

If you decide to grow potatoes, you would need equipment for planting, and lifting the crop, and for spraying the growing crop against disease organisms and their insect vectors. In your first year it would be advisable to pay a contractor to do these jobs for you, with the idea that you would purchase your own equipment later, if the venture warranted it. Clod-formation and stones can

be a problem on some land. Once lifted, the crop should be allowed
to dry off before storing, or bagging. Potatoes must be stored in
the dark; otherwise they turn green. An initial 'curing' period
of about fourteen days, when the temperature is not allowed to
fall below 10-12°C, is important - potatoes are alive and this initial
warm storage period permits them to repair wounds inflicted
during lifting. If potatoes are stored indoors at low temperatures
they will sweeten - this process can be reversed to some extent
by warmer storage conditions for a time prior to marketing them.
An alternative to indoor storage, is to make a clamp outside,
where the potatoes are covered by a plastic sheet with sufficient
earth over it to keep out the frost. A few farmers simply leave
their potatoes in the ground, lifting them as required - they take
the risk that the frost will not reach them, and they cannot harvest
them when the ground is frozen.

If you grow potatoes and sell them to a wholesale distributor, you
will not find your venture very profitable. If, on the other hand, you
have a large van or pick-up truck, and you are prepared to travel
around knocking on peoples' doors, you should find it easy to
retail them in 25 kilogram bags. They are quite easy to sell this
way, for the customer is saved the trouble of carrying the bags, if
he or she went out to buy them. A crop from a few acres, sold
thus, with the benefit of retail price, could bridge the income
gap on many smallholdings.

Surplus and damaged potatoes can be used raw, as stock feed for
older cattle or sheep. They are better fed chipped. Avoid feeding
frozen or rotten potatoes, and those which have turned green;
the latter are poisonous. Raw potatoes can be fed to ewes provided
their diet is changed over gradually; they are better fed to cattle,
as part of their diet only. Potatoes are generally boiled for pigs; for
poultry they should be cooked to a soft mash, to replace part of
their grain ration.

Christmas trees.

Norway spruce may be planted in odd fenced-off corners, on rough
land, or on steep slopes. They may be planted at up to 4000 per
acre. It takes several years to obtain a saleable crop; during this
period the grass and other weeds growing around the bases of the
trees must be cut away, at leat twice a year, or the trees will not
grow to a Christmas-tree shape. Harvesting can last a few years,
the first trees harvested thinning out the crop, the last providing
a Christmas tree from the top, and a fencing stake, plus perhaps
some firewood, from the lower part. Unless you are thinking of
just selling a few trees locally, you should plan your operation

with a view to producing sufficient trees per year to fill at least one lorry - dealers may not wish to buy from you if you have only a part-load, because of their transport costs.

Poultry and rabbits.

We have already treated poultry in detail in chapter 6, and rabbits briefy in chapter 7. Free-range egg, or semi-intensive poultry or rabbit meat production could be a way of closing a smallholding's income gap. At the risk of repetition, we remind you that not all poultry ventures are successful, and it is wisest to start small, keeping a careful watch on your costs, and exploring the market before expanding gradually. Remember that intensive poultry keepers buy their feed in bulk, at a discounted price. You must concentrate, therefore, on the quality of your product, to compensate for the higher price you will have to pay for feed. At the risk of further repetition, we would stress that successful rabbit breeders put a great deal of time and effort into their enterprises, and that a rabbit venture may lose money if the stock are not supervised adequately, especially when they are about to give birth.

Fish ponds.

Setting up a full-scale trout hatchery and farm is outside the scope, and title, of this book. Given suitable conditions on a holding, a fish pond may be made for rearing bought-in fingerling fish from a hatchery, for sale, or home consumption. For trout, the ideal is a good and constant supply of spring water, preferably calcareous, for they are said to grow more slowly in acid waters. Stream or river water is less suitable, because it may bring disease to the fish. A pond should be dug in a situation to suit the water supply, and of a shape to suit the contours of the land. It should be deep enough throughout to discourage wading birds such as herons. Screens should be fixed at the water-exit to prevent fish from escaping. You would have to buy feed for the trout, but they are efficient converters. When you had fish for sale, you could probably advertise for customers to come and collect them, for they would have the enjoyment of the excursion and the satisfaction of really fresh fish. If you have no good supply of spring water, but an area of wet land, you may be able to dig a carp pond. You could stock it with fingerling carp from a hatchery, grow them to a good size, and sell them alive to angling societies, or perhaps to Chinese restaurants. Having a large pond dug is quite expensive, so the annual return from a fish pond is likely to be small in relation to the initial capital outlay; it has to be looked on as a long-term investment.

Crafts, skills, and services.

There is a whole range of activities that may be based on the farm, with a view to closing the income gap. Some of these may be directly linked with farming; others may be linked with skills you acquired in previous employment, or from a hobby. We make a few suggestions; you will be able to think of others yourself.

Agricultural services. If you have the capital, you could think of buying a comprehensive range of machinery and implements; if you are fit and active you could do contract work for other farmers; if not, you could hire out the equipment. Apart from the usual range of equipment, there are specialized items you could hire out, such as ultrasonic scanners for detecting pregnancy, and distinguishing ewes carrying single, twin, and triple lambs. This service is useful for large flock owners, because ewes carrying more than one lamb should be given larger rations. If you are young and active, but short of capital, you could make a start by advertising for work such as erecting fences, digging the post-holes by pick and spade, or with a petrol-driven, or tractor-operated post-hole digger, if you could afford one. Welding is another possibility - we used to know someone in Hampshire who made a living driving around farms with a gas welding kit in the back of a van. If you have a background in clerical work, you could try offering a farmers' secretarial service, doing book-keeping and V A T returns for a modest fee. Many farmers have an aversion to paper work; they are also reluctant to pay the high prices charged by professional accountants. If you have qualifications in chemistry, you could consider setting up a small soil analysis laboratory, perhaps seeking to be paid by one or more fertilizer firms who offer free analysis to farmers in order to obtain their custom. You should check out the possibilities first, before purchasing the appropriate chemicals and equipment. If you are good at woodwork, and can obtain sawn timber cheaply from a sawmill, there is a whole range of items to be made for sale: feeding troughs, hurdles, brooders, sheds, etc.

Non-agricultural arts, crafts, and skills. A husband and wife team between them may be able to produce items at home for sale, or to offer special skills on a self-employed basis. High quality products of the culinary arts, such as gateaux, real French glazed apple tarts, and home-made pâté, could be produced to sell on a regular basis to hotels and restaurants. Many country pubs that only sell main-course bar meals such as chicken, or scampi and chips, could become regular customers for ready-made deserts which they could sell, but would not have the time to make themselves. Sweaters, hand-knitted, or machine-knitted from high quality woollen yarns, hand-woven carpets, pottery, toys, dolls, hand-made furniture, can all be produced at home by those with the appropriate skills. To obtain the right price, hand-made goods generally need to be sold through a tourist outlet, or a retail outlet in a city, which means transport costs and a retailer's cut. If you have professional or trade skills, ranging from accountancy, through building and carpentry, to plumbing and wiring, you can advertise with a view to picking up work from time to time.

Whatever you decide to do, first research the market to ensure there are real possibilities, and then stick to one thing. If you try too many things, perhaps none of them will ever take off. Many of the ideas we have put forward in this chapter will not produce instant returns; those who have sunk all their capital in their holding, and have no other source of income, would be well-advised first to seek part-time, or casual employment, with a view to developing a home-based activity later, when circumstances permit.

Tourism.

If your holding is in an appropriate situation, additional income may be obtained from tourism. The obvious options are bed and break-fast, self-catering accommodation, caravans and tents. Self-catering flats or flatlets may be made in part of the farm-house, space permitting, or in converted outbuildings. Planning permission will be required for this, as also for regular caravan siting. You may, however, keep a caravan 'within the curtilage of your home' without planning permission; you may use it as an extension to your home, but not as an independent dwelling. Likewise you may have caravans and tents on your land for twenty-eight days a year. By making special arrangements with the Camping Club of Great Britain and Northern Ireland, you may have this facility extended to fifty-six days. In many country districts the tourist season is short, little longer than the school summer holidays.

FURTHER READING*.

GENERAL

The Observer's Book of Farm Animals by Lawrence Alderson (Warne) is useful as a beginner's guide to the recognition of different breeds.

The Complete Book of Raising Livestock and Poultry, edited by Katie Thear and Dr. Alistair Fraser is a well-illustrated comprehensive smallholder's guide.

GRASSLAND

Grass Farming by M. McG. Cooper and D. W. Morris (Farming Press) is a readable book for those who wish to go into this subject in more detail.

SHEEP

Profitable Sheep Farming by M. McG. Cooper and R. J. Thomas (Farming Press) is clearly expressed and informative.

The Shepherd's Calendar by Val Stephensen (A. P. P. Cheltenham House, 616 High St., Banbury, Oxon.) is also useful.

CATTLE

Beef Management and Production by Derek H. Goodwin (Hutchinson) is for those thinking of rearing beef cattle.

The Backyard Dairy Book by Len Street (Prism Press) is a well-written guide for those thinking of keeping a house cow.

POULTRY

The Complete Handbook of Poultry-Keeping by Stuart Banks (Ward Lock) is for the serious poultry-keeper.

ANIMAL HEALTH

The TV Vet Books (Farming Press). This series is informative and well-illustrated.

If you have difficulty in obtaining books locally, the following** issue mail-order lists:

Landsman's Bookshop Ltd., Bucker Hill, Bromyard, Hereford.
Smallholder Bookshop, P. O. Box 15, Stowmarket, Suffolk. 1P141PQ.
Small Scale Supplies, Widdington, Saffron Walden, Essex.
Brinsea Products Ltd., Unit 10, Knightcott Estate, Banwell, Avon. BS246JN. (Poultry books only).
S.P.R. Poultry Centre, Barnham Station Park, Barnham, Nr. Bognor Regis, Sussex. (Poultry and self-sufficiency).
D. E. Bonnett, Longcroft, Dowsett Lane, Ramsden Heath, Billericay, Essex.(Poultry only).

*Note this is a personal selection only; there may be other, equally suitable books.
 **We apologise to any suppliers we may have overlooked.

Addendum to page 17, paragraph 4.
The composition of calcified seaweed is variable.
It generally contains only small quantities of phos-
phate and potash, and a range of trace elements.
Its value as a fertilizer is limited and its best use
is as a magnesium-rich liming material.